Prentice-Hall
Series in Automatic Computation

George Forsythe, editor

AHO, editor, *Currents in the Theory of Computing*
AHO AND ULLMAN, *Theory of Parsing, Translation, and Compiling, Volume I: Parsing;*
 Volume II: Compiling
ANDREE, *Computer Programming: Techniques, Analysis, and Mathematics*
ANSELONE, *Collectively Compact Operator Approximation Theory*
 and Applications to Integral Equations
ARBIB, *Theories of Abstract Automata*
BATES AND DOUGLAS, *Programming Language/One,* 2nd ed.
BLUMENTHAL, *Management Information Systems*
BRENT, *Algorithms for Minimization without Derivatives*
BRINCH HANSEN, *Operating System Principles*
COFFMAN AND DENNING, *Operating Systems Theory*
CRESS, et al., *FORTRAN IV with WATFOR and WATFIV*
DAHLQUIST, et al., *Numerical Methods*
DANIEL, *The Approximate Minimization of Functionals*
DEO, *Graph Theory with Applications to Engineering and Computer Science*
DESMONDE, *Computers and Their Uses,* 2nd ed.
DESMONDE, *Real-Time Data Processing Systems*
DRUMMOND, *Evaluation and Measurement Techniques for Digital Computer*
 Systems
EVANS, et al., *Simulation Using Digital Computers*
FIKE, *Computer Evaluation of Mathematical Functions*
FIKE, *PL/1 for Scientific Programmers*
FORSYTHE AND MOLER, *Computer Solution of Linear Algebraic Systems*
GAUTHIER AND PONTO, *Designing Systems Programs*
GEAR, *Numerical Initial Value Problems in Ordinary Differential Equations*
GOLDEN, *FORTRAN IV Programming and Computing*
GOLDEN AND LEICHUS, *IBM/360 Programming and Computing*
GORDON, *System Simulation*
HARTMANIS AND STEARNS, *Algebraic Structure Theory of Sequential Machines*
HULL, *Introduction to Computing*
JACOBY, et al., *Iterative Methods for Nonlinear Optimization Problems*
JOHNSON, *System Structure in Data, Programs, and Computers*
KANTER, *The Computer and the Executive*
KIVIAT, et al., *The SIMSCRIPT II Programming Language*
LAWSON AND HANSON, *Solving Least Squares Problems*
LORIN, *Parallelism in Hardware and Software: Real and Apparent Concurrency*
LOUDEN AND LEDIN, *Programming the IBM 1130,* 2nd ed.
MARTIN, *Design of Man–Computer Dialogues*
MARTIN, *Design of Real-Time Computer Systems*
MARTIN, *Future Developments in Telecommunications*
MARTIN, *Programming Real-Time Computing Systems*
MARTIN, *Security Accuracy and Privacy in Computer Systems*
MARTIN, *Systems Analysis for Data Transmission*
MARTIN, *Telecommunications and the Computer*

COMPUTER APPROACHES
TO
MATHEMATICAL PROBLEMS

MARTIN, *Teleprocessing Network Organization*
MARTIN AND NORMAN, *The Computerized Society*
MATHISON AND WALKER, *Computers and Telecommunications: Issues in Public Policy*
MCKEEMAN, et al., *A Compiler Generator*
MEYERS, *Time-Sharing Computation in the Social Sciences*
MINSKY, *Computation: Finite and Infinite Machines*
NIEVERGELT, et al., *Computer Approaches to Mathematical Problems*
PLANE AND MCMILLAN, *Discrete Optimization: Integer Programming and Network Analysis for Management Decisions*
PRITSKER AND KIVIAT, *Simulation with GASP II: a FORTRAN-Based Simulation Language*
PYLYSHYN, editor, *Perspectives on the Computer Revolution*
RICH, *International Sorting Methods Illustrated with PL/1 Programs*
RUSTIN, editor, *Algorithm Specification*
RUSTIN, editor, *Computer Networks*
RUSTIN, editor, *Data Base Systems*
RUSTIN, editor, *Debugging Techniques in Large Systems*
RUSTIN, editor, *Design and Optimization of Compilers*
RUSTIN, editor, *Formal Semantics of Programming Languages*
SACKMAN AND CITRENBAUM, editors, *On-Line Planning: Towards Creative Problem-Solving*
SALTON, editor, *The SMART Retrieval System: Experiments in Automatic Document Processing*
SAMMET, *Programming Languages: History and Fundamentals*
SCHAEFER, *A Mathematical Theory of Global Program Optimization*
SCHULTZ, *Spline Analysis*
SCHWARZ, et al., *Numerical Analysis of Symmetric Matrices*
SHERMAN, *Techniques in Computer Programming*
SIMON AND SIKLOSSY, editors, *Representation and Meaning: Experiments with Information Processing Systems*
STERBENZ, *Floating-Point Computation*
STERLING AND POLLACK, *Introduction to Statistical Data Processing*
STOUTEMYER, *PL/1 Programming for Engineering and Science*
STRANG AND FIX, *An Analysis of the Finite Element Method*
STROUD, *Approximate Calculation of Multiple Integrals*
TAVISS, editor, *The Computer Impact*
TRAUB, *Iterative Methods for the Solution of Polynomial Equations*
UHR, *Pattern Recognition, Learning, and Thought: Computer-Programmed Models of Higher Mental Processes*
VAN TASSEL, *Computer Security Management*
VARGA, *Matrix Iterative Analysis*
WAITE, *Implementing Software for Non-Numeric Application*
WILKINSON, *Rounding Errors in Algebraic Processes*
WIRTH, *Systematic Programming: An Introduction*

COMPUTER APPROACHES
TO
MATHEMATICAL PROBLEMS

JURG NIEVERGELT

Department of Computer Science
University of Illinois

J. CRAIG FARRAR

Department of Computer Science
Ohio University

EDWARD M. REINGOLD

Department of Computer Science
University of Illinois

PRENTICE-HALL, INC.

ENGLEWOOD CLIFFS, NEW JERSEY

Library of Congress Cataloging in Publication Data

NIEVERGELT, JÜRG.
 Computer approaches to mathematical problems.

 (Prentice-Hall series in automatic computation)
 1. Electronic digital computers—Programming.
2. Electronic data processing—Mathematics.
I. Farrar, J. Craig., joint author. II. Reingold,
Edward M., joint author. III. Title.
QA76.6.N54   519.4   73–3425
ISBN 0–13–164855–1

10   9   8   7   6   5   4   3   2   1

Printed in the United States of America

PRENTICE-HALL INTERNATIONAL, INC., *London*
PRENTICE-HALL OF AUSTRALIA, PTY. LTD., *Sydney*
PRENTICE-HALL OF CANADA, LTD., *Toronto*
PRENTICE-HALL OF INDIA PRIVATE LIMITED, *New Delhi*
PRENTICE-HALL OF JAPAN, INC., *Tokyo*

CONTENTS

5 COMPUTING WITH NUMBERS 174

6 WHAT MACHINES CAN AND CANNOT DO 224

PREFACE

This book is intended as a text for an intermediate course in computer science or mathematics which focuses on the relation between computers and mathematics in formulating and solving problems. In such a course a wide variety of problems would be discussed, the underlying connection being that all of the problems require mathematical concepts for their formulation, and computer techniques for their practical solution.

Two considerations have convinced us that it is time to emphasize, in mathematics and computer science curricula, computer-aided problem formulation and solution. First, over the past two decades since computers have come into general use, a large body of novel concepts and techniques has been developed which has been proven to be of general applicability. We have in mind such notions as random number generation, backtrack, heuristic search, syntax directed processing of expressions, and so on. Some familiarity with these techniques is essential not only for a computer scientist, but also for a sophisticated computer user, or indeed for anybody who wishes to understand how computers accomplish many difficult tasks.

Second, mathematics curricula are changing. It has become increasingly apparent that we have been educating too many research mathematicians and not enough users of mathematics. A user of mathematics formulates problems, wherever they may come from, in mathematical terms, and attempts to solve them by whatever means are available. Today, this frequently implies that a computer will be involved at some stage during the process of formulating and solving a problem. When one deals with problems as complex as those in real-life situations, much insight can often be gained only by means of long computations.

This book has two major goals. Primarily, we want to provide, in one easily accessible source, an elementary presentation of some important practical techniques which have evolved in computer science. We consider this to be particularly important since the literature on these subjects is scattered over many technical publications, and, in addition, is often of a highly technical nature. For the reader who wishes to study some subject in depth, each chapter contains a section called "Remarks and References," which is intended as an entry point into the specialized literature. In addition to this primary goal, we want to show, in numerous examples, how problems are attacked with the help of a computer: how they are formulated mathematically, how algorithms are designed, how computer results are interpreted, and how conjectures are formulated and tested through further computation. We know of no way of teaching these problem solving skills other than by presenting examples.

The chapters of this book are largely self-contained and independent of each other; hence they can be read in any order. By focusing on a few important ideas in each subject discussed, the book aims at being a framework around which a teacher can design an interesting course which reflects strongly his own views and preferences.

It is not easy to state formal prerequisites for this book. It assumes a familiarity with and experience in computer programming which can be obtained in a first course on computers, one in which students write several rather extensive programs. This background should suffice to write programs for the algorithms that are discussed. The book also assumes a mathematical maturity typical of a junior or senior mathematics major, and a willingness to accept, as plausible, an intuitive argument where a formal proof lies outside the scope of this book.

This book grew out of a second course in computer science developed at the University of Illinois for secondary school mathematics teachers. The goal of the course was to provide teachers with a collection of problems from diverse areas of mathematics which lend themselves to a computer approach, and to show how such problems are attacked with the help of a computer. The intention was that the teachers would feel confident to adapt these problems to the high school level and thus to be able to illustrate the use of computers in their own teaching. As the course evolved and began to attract other students in computer science and in mathematics, we became convinced that the audience for such a course was much larger than was originally expected. If we had to describe this audience in one sentence, we would say it consists of those people who are interested in the relationship between computer science and mathematics.

ACKNOWLEDGMENTS

Our thanks are due to the many students at the University of Illinois who have contributed indirectly, but substantially, to this book. The feedback we have received while teaching this material over the past several years has been essential for the choice of topics, for the depth to which a topic is pursued, and for the style of presentation. By showing us that these topics were well-received, it has also given us the encouragement necessary to complete the lengthy task of writing this book.

We are grateful to the Association for Computing Machinery for permission to reproduce the article "What Machines Can and Cannot Do" (J. Nievergelt and J. C. Farrar. *Computing Surveys*, *4*, 2 (June 1972), 81–96). Chapter 6 is a slightly modified version of this article.

Many helpful comments were provided by Richard V. Andree, Stephen M. Chase, Paul Chouinard, Dennis F. Cudia. Michael J. Fischer, Robert Foster, T. Kameda, John D. Lipson, Mateti Prabhaker, Jean L. Pradels, Martin Schultz, and Thomas R. Wilcox.

We would be remiss if we did not also thank Mrs. Connie Slovak for her excellent typing and retyping of this manuscript.

JURG NIEVERGELT

J. CRAIG FARRAR

EDWARD M. REINGOLD

1 WHAT IS THERE TO ARITHMETIC EXPRESSIONS ?

Mathematicians tend to stress the importance of concepts and ideas, and to treat the notation in which these ideas are expressed as a matter of mere convenience, devoid of any intrinsic interest; however, there are several arguments that make a strong case for the importance of the notation per se. One is the historical evidence that various branches of mathematics have made great advances as a consequence of, or in conjunction with, the invention of an appropriate notation; a case in point is arithmetic, which, after the development of Arabic numerals, changed from a poorly understood art to the common knowledge of every educated person. Another argument for the study of notation is the investigation into the foundations of mathematics which began around the turn of this century, largely in response to contradictions that arose in set theory. Logicians found it necessary to develop *formal* languages, whose interpretation was not left to the intuition of the reader, but could be described in a rigorous, "mechanical" way. Any proof or derivation written in these languages must be "effective," in the sense that its correctness can be checked automatically by some machine.

The latter remark leads us to explain our concern with notation. It is usually desirable to give instructions to a computer in a language that allows no ambiguity. It is difficult, although not necessarily impossible, for a computer to extract any reasonable information from a statement like "Well, you know what I mean." Most programming languages attempt to be formal, in the sense that it is absolutely clear what statements can be made in this language, and that it is absolutely clear how statements are to be interpreted.

This chapter presents two formal languages for arithmetic expressions and shows how the precision gained through this approach allows one to state clearly and concisely some important algorithms on arithmetic expressions.

1.1. CONVENTIONAL PARENTHESIS NOTATION

An arithmetic expression is a sequence of characters that might look like

$$(R + H)/2 * 3.14.$$

One learns how to write and use them by example, without ever needing an exact definition of what an arithmetic expression is or what it denotes. When presenting arithmetic expressions to a computer, however, it is necessary first to describe precisely to the computer the class of all arithmetic expressions that may ever be encountered, and then to tell the computer precisely what is to be done with each one of them. Since in all interesting cases this class is infinite, we cannot prepare an exhaustive dictionary; rather, we must give to the computer a finite definition of an infinite class. This often takes the form of a list of explicitly presented, very simple arithmetic expressions, and some *rules of inference*, which say that if such-and-such are arithmetic expressions, then some combination of them is again an arithmetic expression.

For a simple formal system such as the "language" of arithmetic expressions, we shall also insist that it be possible, given any string of characters, to effectively and unambiguously answer the question of whether it is or is not an arithmetic expression, so that incorrect expressions can be automatically rejected (see Exercise 2). Once we have achieved this, we know how arithmetic expressions "look," but we still have to say how they are to be interpreted. Even though the word "interpreted" can mean many things, as we shall see later, some aspects of the meaning of arithmetic expressions can be discussed now. For example, if

$$(R + H)/2 * 3.14$$

turns out to be an arithmetic expression in our formal language, we have to say whether it means

$$((R + H)/2) * 3.14$$

or

$$(R + H)/(2 * 3.14).$$

These two aspects of arithmetic expressions, their appearance and their meaning, are referred to respectively as the *syntax* and the *semantics* of the language of arithmetic expressions. In a formal language such as we shall develop both must be rigorously defined.

A definition of a formal language begins by stating what its alphabet is or, in other words, what characters or symbols may be used. Our formal language for arithmetic expressions will have the following characters:

1. The letters A through Z
2. The decimal digits 0 through 9
3. The operators $+ - * /$
4. Three remaining characters . ( )

The following rules, called *productions*, serve to define both the alphabet and its partition into these four classes.

$$\langle\text{character}\rangle :: = \langle\text{letter}\rangle \mid \langle\text{digit}\rangle \mid \langle\text{operator}\rangle \mid$$
$$\langle\text{remaining character}\rangle$$
$$\langle\text{letter}\rangle :: = A \mid B \mid C \mid D \mid E \mid F \mid G \mid H \mid I \mid J \mid K \mid L \mid M \mid N \mid$$
$$O \mid P \mid Q \mid R \mid S \mid T \mid U \mid V \mid W \mid X \mid Y \mid Z$$
$$\langle\text{digit}\rangle :: = 0 \mid 1 \mid 2 \mid 3 \mid 4 \mid 5 \mid 6 \mid 7 \mid 8 \mid 9$$
$$\langle\text{operator}\rangle :: = \langle\text{A-operator}\rangle \mid \langle\text{M-operator}\rangle$$
$$\langle\text{A-operator}\rangle :: = + \mid -$$
$$\langle\text{M-operator}\rangle :: = * \mid /$$
$$\langle\text{remaining character}\rangle :: = . \mid (\mid)$$

These productions are read as follows:

A character of the language to be defined may be a letter or a digit, or an operator, or a remaining character. A letter may be an A, or a B, or a C, or . . . , or a Z. A digit may be a 0, or a 1, or a 2, and so on.

The reason for dividing the class of operators into two subclasses, $+$ and $-$ on the one hand and $*$ and $/$ on the other, should become apparent later in this section. Additive operators and multiplicative operators play different roles in our conventional notation for arithmetic expressions.

Notice that the vertical line that separates alternatives on the right-hand side of a production, the symbol $:: =$ which we read as "may be any one of," and the angular brackets are not part of the language of arithmetic expressions we are defining. They are, instead, part of the metalanguage in which we write the definition, English augmented by some mathematical formalism. Later it will be useful to consider the words enclosed in angular brackets, such as $\langle\text{operator}\rangle$, as single symbols in this metalanguage. Hence we call them *metasymbols*, or *nonterminal symbols*, a name that comes from the way we shall use these symbols later.

It would be conceptually simpler if we used a different typesetting, say boldface, for characters in the language we want to define, but when this is not possible, we must resort to a convention like the one adopted here; that is, any character enclosed in angular brackets is not to be considered a character of the formal language being defined; this works if the angular brackets themselves do not belong to the alphabet of this language.

Next, we define two important elements of arithmetic expressions, constants and variables:

$$\langle\text{constant}\rangle::=\langle\text{digit}\rangle\langle\text{digit*}\rangle\,|\,\langle\text{digit*}\rangle\cdot\langle\text{digit}\rangle\langle\text{digit*}\rangle$$
$$\langle\text{variable}\rangle::=\langle\text{letter}\rangle\langle\text{letter*}\rangle$$

Here we have introduced some new notation: by $\langle\text{digit*}\rangle$ and $\langle\text{letter*}\rangle$ we mean "any finite string of digits" and "any finite string of letters," respectively; in both cases we include the *null string*, which has no characters at all. For example, '7' and '1968' are character strings denoted by $\langle\text{digit*}\rangle$, and 'X' and 'RADIUS' are character strings denoted by $\langle\text{letter*}\rangle$.

Thus these last two productions mean that a constant is a sequence of digits that may or may not contain a decimal point (if it does, the decimal point may occur anywhere except at the right end). Hence '007', '.5', and '3.14159' are character strings denoted by $\langle\text{constant}\rangle$. A variable is a string of letters; for example, 'CAT', 'DOG', and 'X' are character strings denoted by $\langle\text{variable}\rangle$.

Whenever ambiguities might arise as to where a specific character string begins and ends, we shall enclose this string in single quotes as was done in the previous examples. This notation allows us to make distinctions between strings like 'A' and 'A ', the latter being a string of two characters, 'A' followed by a blank, and it also allows us to represent the null string, which has no characters, by ' '. Whenever we write an arithmetic expression in single quotes, we shall use no blanks, since we chose not to allow the blank in the alphabet of the formal language of expressions.

The next three productions are the crux of our scheme. These productions cannot be described separately, as the definition of each involves the others. Such definitions are called *recursive*, and they play an important part in certain branches of mathematics, such as logic, and in computer programming.

$$\langle\text{factor}\rangle::=\langle\text{constant}\rangle\,|\,\langle\text{variable}\rangle\,|$$
$$(\langle\text{arithmetic expression}\rangle)$$
$$\langle\text{term}\rangle::=\langle\text{factor}\rangle\,|\,\langle\text{term}\rangle\langle\text{M-operator}\rangle\langle\text{factor}\rangle$$
$$\langle\text{arithmetic expression}\rangle::=\langle\text{term}\rangle\,|$$
$$\langle\text{arithmetic expression}\rangle\langle\text{A-operator}\rangle\langle\text{term}\rangle$$

To understand what these productions mean, consider for a moment an extended class of expressions which can be written in an alphabet that consists of the characters of the language of arithmetic expressions, plus the meta-symbols, that is, the words enclosed in angular brackets. To avoid confusion with arithmetic expressions, let us call such extended expressions meta-expressions. The productions can then be considered as rules for rewriting metaexpressions, one metasymbol at a time, as in the following example. By applying the production

$$\langle\text{arithmetic expression}\rangle :: = \langle\text{arithmetic expression}\rangle \langle\text{A-operator}\rangle \langle\text{term}\rangle$$

as a rewriting rule, one obtains from the metaexpression

$$\langle\text{arithmetic expression}\rangle + \langle\text{term}\rangle$$

a metaexpression

$$\langle\text{arithmetic expression}\rangle \langle\text{A-operator}\rangle \langle\text{term}\rangle + \langle\text{term}\rangle;$$

or by applying the production

$$\langle\text{term}\rangle :: = \langle\text{term}\rangle \langle\text{M-operator}\rangle \langle\text{factor}\rangle,$$

one obtains

$$\langle\text{arithmetic expression}\rangle + \langle\text{term}\rangle \langle\text{M-operator}\rangle \langle\text{factor}\rangle.$$

In this situation we say that the second metaexpression is immediately derived from the first by an application of the given production. Generally we say that a metaexpression E_{n+1} is derived (or has a derivation) from metaexpression E_1 if there is a sequence of metaexpressions

$$E_2, E_3, \ldots, E_n$$

and a sequence of productions

$$P_1, P_2, \ldots, P_n$$

such that for each $i = 1, 2, \ldots, n$, E_{i+1} is immediately derived from E_i by an application of P_i.

We see now that the conventional name *nonterminal symbol* for a metasymbol comes from the fact that these symbols can be replaced, or rewritten, during a derivation. By contrast, a character of the language of arithmetic expressions cannot be rewritten, and hence it is also called a *terminal symbol*. In these terms, then, the set of arithmetic expressions defined by the productions can be defined as follows: an arithmetic expression is a string of terminal symbols that has a derivation from the metaexpression

$$\langle\text{arithmetic expression}\rangle,$$

using the given productions. The following derivation for 'A/(B + C)' may help to clarify these ideas.

```
                    ⟨arithmetic expression⟩
                        ⟨term⟩
        ⟨term⟩  ⟨M-operator⟩  ⟨factor⟩
       ⟨factor⟩  ⟨M-operator⟩  ⟨factor⟩
     ⟨variable⟩  ⟨M-operator⟩  ⟨factor⟩
       ⟨letter⟩  ⟨M-operator⟩  ⟨factor⟩
            A    ⟨M-operator⟩  ⟨factor⟩
            A    /             ⟨factor⟩
            A    /             (⟨arithmetic expression⟩)
            A    /   (⟨arithmetic expression⟩  ⟨A-operator⟩  ⟨term⟩)
            A    /   (⟨term⟩               ⟨A-operator⟩  ⟨term⟩)
            A    /   (⟨factor⟩             ⟨A-operator⟩  ⟨term⟩)
            A    /   (⟨variable⟩           ⟨A-operator⟩  ⟨term⟩)
            A    /   (⟨letter⟩             ⟨A-operator⟩  ⟨term⟩)
            A    /   (   B                 ⟨A-operator⟩  ⟨term⟩)
            A    /   (   B                     +         ⟨term⟩)
            A    /   (   B                     +         ⟨factor⟩)
            A    /   (   B                     +         ⟨variable⟩)
            A    /   (   B                     +         ⟨letter⟩)
            A    /   (   B                     +         C      )
```

Our intention in setting up these productions is to consider an arithmetic expression as a sequence of terms, with each two consecutive terms being separated by an A-operator, '+' or '−'. Thus in the character string

$$\text{`3}-X+2*Y-Z*(A+B-C*D)/(X-Y)\text{'},$$

which is an arithmetic expression, the substrings

$$\text{`3'},\quad \text{`X'},\quad \text{`2*Y'}\quad \text{and}\quad \text{`Z*(A+B-C*D)/(X-Y)'}$$

are terms.

A term is a sequence of factors, each two consecutive factors being separated by an M-operator, '*' or '/'. In the example, the terms '3' and 'X' each consist of a single factor, while '2*Y' has two factors, '2' and 'Y', and the last term has the three factors 'Z', '(A+B−C*D)', and '(X−Y)'. Notice that an arbitrary arithmetic expression can be a factor, provided it is enclosed in parentheses; this shows the recursive nature of our definition. The arithmetic expression 'A+B−C*D', which, when enclosed in parentheses, entered as a factor into the larger arithmetic expressions discussed earlier, has three terms, 'A', 'B', and 'C*D', the first two of which are also factors, while the third has the two factors 'C' and 'D'.

A set of productions such as we have given is called a *grammar*. The language of arithmetic expressions can be defined by other grammars, some of which are "simpler." However, our choice has not been arbitrary; the productions we have chosen generate arithmetic expressions in accordance with the usual convention by which multiplication and division take precedence over addition and subtraction in their evaluation; the important consequence of this choice is that the syntactical structure is made to describe part of the semantics (see also Exercise 3).

A word about the nonterminal symbols ⟨character⟩ and ⟨remaining character⟩ is in order, since they can never appear in a derivation. They are introduced to list explicitly the alphabet of the language of arithmetic expressions.

The arithmetic expressions defined here are admittedly more restricted than is usual; for example, '−7' is not an arithmetic expression according to the definition. It would be easy to incorporate productions that allow a unary + or − to occur in an expression (see Exercise 4); we have not done so since this would complicate the description without adding any insight.

The syntactical structure of an arithmetic expression may be visualized by means of a convenient graphical device, called a *parsing* or *derivation tree*, which is used throughout this chapter. A production of the form

$$\langle S \rangle ::= S_1 \, S_2 \, S_3 \ldots S_n$$

where each S_i on the right is either a string of characters in the alphabet or a word enclosed in angular brackets corresponds to the tree

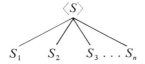

It is really just a trivial rewriting of the production, as can be seen from Figure 1.1, which illustrates the trees for several of the productions.

As we stated previously, a string of characters from the alphabet is an arithmetic expression only if it can be derived using the productions. Thus for each expression there is a sequence of the productions whose application defines the expression, and the derivation in terms of these productions determines a tree, as shown in Figure 1.2, for the expression '(R+H)/2*3.14'. The root of this tree bears the label ⟨arithmetic expression⟩, and each of its leaves is labeled with a single character of the language. Furthermore, it has the property that each nonterminal node, the branches leaving it, and the nodes connected thereby form a copy of a tree corresponding to a production for the language. Any tree that has these properties is known as a parsing tree or derivation tree for some arithmetic expression, and is a representation of the complete syntactical structure of the expression. The particular expression may be read from left to right in the terminal nodes of the tree. Parsing an

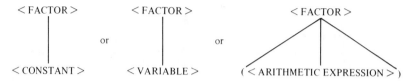

A. The production < factor > : : = < constant > | < variable > | (< arithmetic expression >)

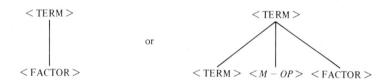

B. The production < term > : : = < factor > | < term > < M-operator > < factor >

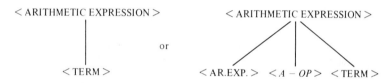

C. The production
< arithmetic expression > : : = < term > | < arithmetic expression > < A-operator > < term >

Fig. 1.1   Tree representation of productions.

expression is the process of describing its syntactical structure or constructing its parsing tree. The reader should try to become proficient at parsing arithmetic expressions, as a knowledge of the parsing tree for an arbitrary expression will be needed later; to this end, see Exercise 1.

The reader should convince himself that the parsing tree of Figure 1.2 is the only one possible for the expression '(R+H)/2*3.14'†; this means, for example, that there is no way to complete a tree that starts out as shown in Figure 1.3. This uniqueness of the parsing tree, which is true for all expressions in the language (Exercise 5), gives a one-to-one correspondence between the

†On the other hand, there are many possible derivations of this expression, because any metaexpression that contains more than one nonterminal symbol can generate different next metaexpressions, depending on which symbol is rewritten. One reason why a parsing tree is an intuitively appealing representation is that it ignores the accidental order in which different nonterminal symbols in the same metaexpression are rewritten, while retaining the essential order showing which symbols derive from which others.

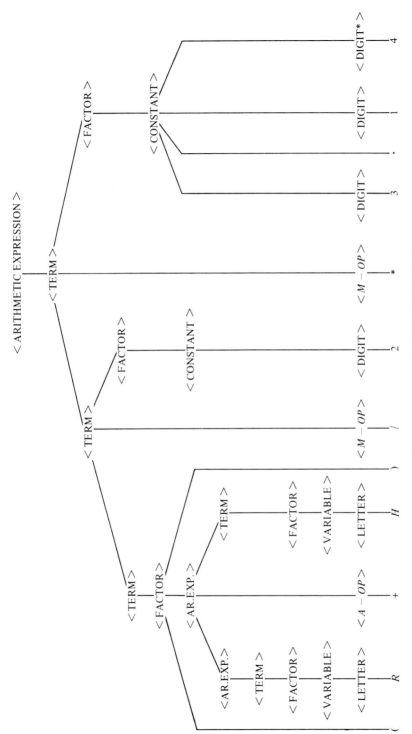

Fig. 1.2 Complete parse tree for '(R + H)/2 * 3.14'.

9

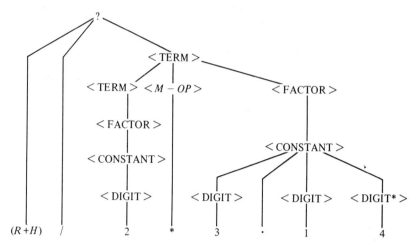

Fig. 1.3 An impossible parse of '(R + H)/2 ∗ 3.14'.

set of arithmetic expressions and the set of parsing trees. It also makes it possible to perform manipulations on expressions indirectly by working with the parsing trees, using as a guide the syntax information contained in the tree. Algorithms that use the syntactical structure of an expression in this way are called *syntax directed*; the algorithms of Section 1.4 are of this nature.

The uniqueness of the parsing tree for an arithmetic expression is not an accident, nor is it a property of formal languages in general. Again, it is a property of the particular set of productions we have used to define the language; a set of productions with this property is called an *unambiguous* grammar (see Exercise 6).

In this section we have seen how a certain class of strings of alphanumeric characters, the familiar arithmetic expressions, can be defined by a set of productions, a grammar. The use of this definition imparts to each expression a unique structure, its syntactical structure, which can be described easily by means of a tree structure, a parsing tree. In later sections we shall exploit the syntactical structure of our arithmetic expressions to define some important algorithms for the manipulation of expressions by a computer, but we shall first discuss a different language of expressions that computer science has found very useful, and which will furnish a basis for defining the other important aspect of arithmetic expressions, their semantics.

1.2. ANOTHER LANGUAGE: POLISH NOTATION

Although the notation of the previous section is widely used, it is inconvenient for many uses in mathematics and computer science; among its disadvantages are two of fundamental importance. First, additional symbols for "grouping" must be introduced to avoid ambiguity; these symbols are usually parentheses or brackets which have no intrinsic meaning, but are used only to indicate the order in which operations are performed. This is complicated further by the use of various conventions for the omission of parentheses. For example, the arithmetic expression 'A∗B+C' is normally interpreted as '(A∗B)+C' rather than 'A∗(B+C)' since the usual convention is for multiplication to take precedence over addition. This may seem like a minor inconvenience, but because of it the evaluation of such arithmetic expressions is quite tedious (see Exercise 12). The second disadvantage, in reality a consequence of the first, is that the introduction of parentheses or brackets results in a loss of uniqueness in the representation of arithmetic expressions. One would like the two strings '(A+B)∗C' and '((A+B)∗C)' to represent the same expression, but using the productions of the previous section each of these expressions has a different parse tree.

To eliminate these disadvantages, we must look carefully at the basic elements of arithmetic expressions: operators and operands. These are easily defined in terms of the productions: an operator is any string that can be derived from ⟨operator⟩, whereas an operand is any string that can be derived from either ⟨constant⟩ or ⟨variable⟩. All the operators allowed by our productions happen to be *binary*; that is, given two inputs (operands) they produce a single output. In the usual notation, called *infix notation*, the operator is placed between the two operands, with parentheses added as needed. There is an alternative; in particular, mathematical notation frequently uses *prefix notation* in which the operator precedes its operands. Thinking in terms of the arithmetic functions of addition, subtraction, multiplication, and division, this may seem strange, but in terms of functions with *names* it is quite common. Consider, for example, the function "min" defined as

$$\min(x, y) = \begin{cases} x & \text{if } x \leq y, \\ y & \text{otherwise.} \end{cases}$$

Here the operator, that is, the function name, is placed before the operands; the same convention is used for functions like max, sin, cos, and log, and in general for an arbitrary function, say f or g. Giving the definition of a function such as

$$f(x, y) = 2^x 3^y,$$

one would be unlikely to use '$x f y$' instead of '$f(x, y)$'. For arithmetic expressions, we can consider the characters $+$, $-$, $\ast$, and $/$ to be names of functions which can be placed before or after the operands, rather than between them; thus we could use '$+xy$' or '$xy+$' instead of '$x+y$'.

These two notations, in which the operator precedes or succeeds the operands rather than separating them, are due to the Polish logician Jan

Łukasiewicz and are usually known as Polish prefix and Polish suffix, respectively. Both of these notations obviate the difficulties outlined above; but since Polish suffix notation is generally more useful in computer science, it will be covered in some depth; Polish prefix notation will be left to the exercises.

Before giving a rigorous definition of the syntax of Polish suffix notation with a sequence of productions, it is worthwhile to discuss the notation on an intuitive level. As mentioned previously, the idea is to have the two operands juxtaposed and followed by an operator. Since in this notation two variable names or constants might be placed adjacent to one another without an operator between them, we must introduce a notational device called a *delimiter*. This delimiter, the character '_', will serve to separate the various components from each other. The following are examples of the Polish suffix representation of some arithmetic expressions:

Expression	Polish Suffix Form
'DOG+CAT'	'DOG_CAT_+'
'((DOG)+(CAT))'	'DOG_CAT_+'
'A*B+C'	'A_B_*_C_+'
'A*(B+C)'	'A_B_C_+_*'

Notice in these examples that the Polish suffix form of an expression requires no parentheses to define the order in which the operations are performed; moreover, the arithmetic expressions '((DOG)+(CAT))' and 'DOG+CAT' have the same Polish suffix form. The delimiter '_' that has been added does not affect the syntax of the suffix expressions; it serves only to distinguish between expressions such as 'A_BC_+' and 'AB_C_+', which are equivalent to 'A+BC' and 'AB+C', respectively.

A formal language for arithmetic expressions in Polish suffix notation can be considerably simpler than for parenthesis notation. If $\langle P \rangle$ denotes "arithmetic expression in Polish suffix notation," then the following productions define the syntax of the language:

$$\langle \text{character} \rangle ::= \langle \text{letter} \rangle | \langle \text{digit} \rangle | \langle \text{operator} \rangle | \langle \text{decimal point} \rangle |$$
$$\langle \text{delimiter} \rangle$$
$$\langle \text{letter} \rangle ::= A | B | C | D | \ldots | Z$$
$$\langle \text{digit} \rangle ::= 0 | 1 | 2 | 3 | 4 | 5 | 6 | 7 | 8 | 9$$
$$\langle \text{operator} \rangle ::= + | - | * | /$$
$$\langle \text{decimal point} \rangle ::= .$$
$$\langle \text{delimiter} \rangle ::= \_$$
$$\langle \text{constant} \rangle ::= \langle \text{digit} \rangle \langle \text{digit*} \rangle | \langle \text{digit*} \rangle . \langle \text{digit} \rangle \langle \text{digit*} \rangle$$
$$\langle \text{variable} \rangle ::= \langle \text{letter} \rangle \langle \text{letter*} \rangle$$
$$\langle P \rangle ::= \langle \text{constant} \rangle | \langle \text{variable} \rangle | \langle P \rangle \_ \langle P \rangle \_ \langle \text{operator} \rangle$$

Note that there are fewer productions than before, and, in particular, the auxiliary concepts ⟨term⟩, ⟨factor⟩, ⟨A-operator⟩, and ⟨M-operator⟩ are not needed. This new set of productions is simpler not because we have rearranged the order of the operator and operands, but rather because we have eliminated the need for parentheses and hence also the arbitrary conventions for deleting some, but not all, parentheses. If we ignored such conventions in the first place and forced all expressions to be completely parenthesized, a grammar for the parenthesis notation could be as simple as the one above.

Figure 1.4 shows the parsing tree for the Polish suffix expression 'ALPHA_BETA_+_2_−_3.14_Z_/_*'.

Given a normal parenthesis expression, we would undoubtedly have little trouble in evaluating it, working on the innermost parenthetical expressions first and the outermost last. However, it is not at all easy to convert the informal prescription "work from inside out" into a precise description of an evaluation algorithm, as can be seen from Exercise 12. On the other hand, suppose that we were given a Polish suffix expression to evaluate. This is a relatively easy task, even though we may be somewhat unfamiliar with the notation; there are no parentheses to keep track of since the order in which the operations are performed is defined implicitly by their relative order in the expression. Consider, for example, the expression 'A_B_+_C_−_5_D_/_*', which corresponds to the expression '((A+B)−C)*(5/D)'. The order of evaluation is as follows:

$$
\begin{array}{l}
\underbrace{A\_B\_+}\_C\_−\_5\_D\_/\_* \\
\underbrace{A+B} \\
\underbrace{(A+B)−C} \\
\qquad\qquad \underbrace{5/D} \\
\underbrace{((A+B)−C)*(5/D)}
\end{array}
$$

The general rule in performing such an evaluation is: whenever we find two operands followed by an operator, we apply the operator to those operands and then replace the entire substring (the operands and operator) by the value so obtained. Thus, in the previous expression we replaced the substring 'A_B_+' by the value of 'A+B'; then we replaced the substring consisting of the value of 'A+B' followed by 'C_−' by the value of '(A+B)−C', and so forth, eventually replacing the entire string with the value of '((A+B)−C)*(5/D)'.

To specify this evaluation algorithm more precisely, we make use of a *push-down stack*. This is a programming device that arises frequently in algorithms for manipulating arithmetic expressions; essentially, a stack is a data structure in which items move in and out according to the following rule:

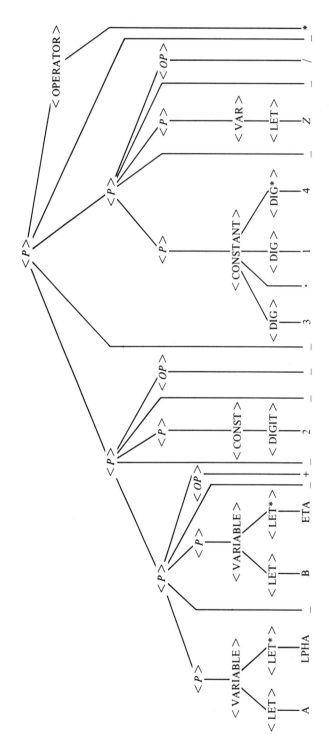

Fig. 1.4 Parse of the Polish suffix expression 'ALPHA_BETA_+_2_−_3.14_Z_/_*'.

when an item comes in it is placed on the top of the stack, and the only item that can be removed from the stack is the top item. Thus the latest entry in is the first out; hence if items A, B, and C entered the stack in that order, they must leave the stack in the order C, B, A.

Using a stack, the algorithm for evaluating Polish suffix expressions scans the character string from left to right, placing operands onto the stack and applying operators to the top two elements of the stack; the result is then placed on top of the stack. More formally, this would be

Step 1

> Read the characters up to the next delimiter, or if there is none, then to the end of the string.

Step 2

> If what was read is an operand, place the value of that operand on top of the stack and return to step 1.

Step 3

> If what was read is an operator, apply that operator to the two topmost elements of the stack, placing the result on top of the stack, and return to step 1.

The following sequence of steps illustrates the use of this algorithm on the input 'A_B_+_C_−_5_D_/_*'; the stack contains the values of the expressions shown:

Input String	Stack Contents		
A_B_+_C_−_5_D_/_*	Empty		
B_+_C_−_5_D_/_*	A		
+_C_−_5_D_/_*	A B		
C_−_5_D_/_*	(A+B)		
−_5_D_/_*	(A+B)	C	
5_D_/_*	((A+B)−C)		
D_/_*	((A+B)−C)	5	
/_*	((A+B)−C)	5	D
*	((A+B)−C)	(5/D)	
Exhausted	(((A+B)−C)*(5/D))		

Because of its simplicity, Polish notation is frequently used to represent expressions in computers. For example, many compilers translate arithmetic expressions from parenthesis notation to Polish suffix as an intermediate step in generating machine language instructions to evaluate the expression. This translation is quite straightforward. We shall describe two algorithms for converting from parenthesis notation to Polish suffix notation.

The first algorithm closely mirrors the syntactic structure of expressions as given by our productions; for example, the recursive nature of the productions is reflected in the recursive nature of the algorithm. It is typical of such *syntax-directed* algorithms that they can be stated concisely and are easy to understand.

Let S be an arithmetic expression, and define $T\langle S \rangle$ as follows:

If S is a constant or a variable, then S.

If S has the form $S_1 \, S_2 \, S_3$, where S_1 is '(', S_3 is ')', and S_2 is an expression, then $T\langle S_2 \rangle$.

If S has the form $S_1 \, S_2 \, S_3$, where S_1 is a term, S_2 is either '*' or '/', and S_3 is a factor, then $T\langle S_1 \rangle$ '_' $T\langle S_3 \rangle$ '_' S_2.

If S has the form $S_1 \, S_2 \, S_3$, where S_1 is an expression, S_2 is either '+' or '−', and S_3 is a term, then $T\langle S_1 \rangle$ '_' $T\langle S_3 \rangle$ '_' S_2.

In some programming languages (e.g., SNOBOL) a program can be written that has exactly the recursive, syntax-directed structure of this algorithm.

The second algorithm for converting from parenthesis to suffix notation looks more complicated and is not as easy to understand, primarily because it uses much simpler primitive operations than the preceding one. For example, it calls only for the recognition of a single variable, constant, operator, or parenthesis, whereas the preceding algorithm assumes that the syntactic structure of an entire expression is given to it. It is presented here to show how the powerful operations demanded by the first algorithm can be expressed in terms of operations that are closer to those available on today's computers.

This algorithm also uses a stack: as the input string is read, operands go directly into the output string, and the stack is used to temporarily store parentheses and operators. Each operator and the open parenthesis has associated with it a priority number; these priorities determine the movement of operators into the stack or into the output, and they are based on the order of evaluation, which is implicit in the productions of Section 1.1: parenthesized expressions first, then multiplications and divisions, and finally additions and subtractions. These priority numbers are

Character	Priority
(	0
+,−	1
*,/	2
Empty stack	0

In more formal terms, the algorithm is given as follows (initially, the stack is assumed to be empty):

Step 1

    If there are no more characters in the input string, go to step 6; otherwise, read the next element (parenthesis, operator, or operand) and call it x.

Step 2

    If x is an operand, put it into the output followed by a delimiter and go back to step 1.

Step 3

    If x is an opening parenthesis, put it on the stack and go to step 1. If x is a closing parenthesis, go to step 4, otherwise go to step 5.

Step 4

    If the top of the stack is not an opening parenthesis, remove it from the stack and put it into the output followed by a delimiter; then repeat step 4. If the top of the stack is an opening parenthesis, remove it from the stack and go back to step 1.

Step 5

    If the priority of the element at the top of the stack is greater than or equal to the priority of x, remove the top element of the stack and put it in the output followed by a delimiter; then repeat step 5. Otherwise, push x onto the top of the stack and go back to step 1.

Step 6

    Empty the stack, one element at a time, each one followed by a delimiter, into the output and we are done.

The following sequence of steps illustrates how this algorithm translates '(A+B*C)/E−F' into 'A_B_C_*_+_E_/_F_−':

Value of x	Input String	Stack Contents	Output String
Empty	(A+B*C)/E−F	Empty	Empty
(	A+B*C)/E−F	(	Empty
A	+B*C)/E−F	(	A_
+	B+C)/E−F	(+	A_
B	*C)/E−F	(+	A_B_
*	C)/E−F	(+*	A_B_
C	)/E−F	(+*	A_B_C_
)	/E−F	(+	A_B_C_*−
)	/E−F	(	A_B_C_*_+_
)	/E−F	Empty	A_B_C_*_+_
/	E−F	/	A_B_C_*_+_
E	−F	/	A_B_C_*_+_E_
−	F	−	A_B_C_*_+_E_/_
F	Empty	−	A_B_C_*_+_E_/_F_
F	Empty	Empty	A_B_C_*_+_E_/_F_−

1.3. SEMANTICS AND EQUIVALENCE

In the preceding sections our discussion has been limited to the syntax, that is, the form of arithmetic expressions, with little mention of what arithmetic expressions stand for. Arithmetic expressions are usually used, however, as names to denote something, and when we operate on expressions, we are really concerned with the corresponding operations on the objects represented by the expressions. What are these objects? What do the expressions really mean? When human beings use and manipulate arithmetic expressions, they usually give little thought to this aspect; they seem to "know," intuitively, what the expressions mean. They rely on experience to guide them so as to operate on expressions only in ways that are consistent with their (unformalized) interpretation of the expressions.

Using a computer frequently forces us to think more explicitly about many operations that we carry out in a habitual manner, and arithmetic expressions provide a good example of this. This is because a computer, of course, cannot know what our interpretation of expressions is unless we describe this in the same precise way we described their syntax. If we fail to describe our interpretation explicitly, some interpretation will be implicit in the program we write, and this may or may not coincide with our intuitive interpretation.

Those aspects of a language which refer to its interpretation, to the meaning of sentences or expressions in that language, are usually called its *semantics*. In this section we shall examine and make precise some notions of the semantics of arithmetic expressions; this, in turn, will allow us to discuss the equivalence of arithmetic expressions: two expressions are equivalent if and only if they have the same meaning.

It should not come as a surprise that we can interpret arithmetic expressions in many different ways, and that these various interpretations are useful in different contexts. Perhaps the most general meaning we can assign to an arithmetic expression is to say that it denotes a function from some domain into some range; the domain and range could be the real numbers, the rationals, and so on. Using this *functional semantics* of expressions has the following consequence: there are many expressions that denote the same function, but have entirely different forms, for example

$$\text{`}(A+B)/(C+C)\text{'},$$
$$\text{`}(A+B)/(2*C)\text{'},$$
$$\text{`}.5*(A+B)/C\text{'}.$$

In this functional semantics, we may choose to either attach importance to or ignore the identity of variables†. In the latter case, an expression like

$$`(X+Y)*(.5/Z)'$$

also denotes the same function.

The preceding interpretation of expressions is not the most useful one for computation. Each of the previous expressions specifies a distinct way of evaluating the same function, but because of roundoff errors (see Section 5.1) or the possibility of overflow, the values actually computed according to these expressions may turn out to be different.

Another definition we could choose for semantics is that each character string which is a legal arithmetic expression in our language would have as its meaning exactly the string itself. Thus every addition of a redundant pair of parentheses would change the meaning of the expression, and so expressions like

$$`(A+B)',$$

$$`A+B',$$

$$`(A)+(B)',$$

$$`((A)+(B))',$$

would each have a different *literal* semantic meaning. This definition is less helpful than the previous one; the former ignores information about the manner of computing the function, but the latter ignores nothing at all, even redundant parentheses.

A compromise between these two definitions for the semantics of arithmetic expressions is Polish suffix notation. As seen in the previous section, this has three important properties: redundant parentheses disappear, making expressions such as '(A+B)' and 'A+B' indistinguishable; nonredundant parentheses that specify the order in which operations are performed are implicitly preserved; and it is easy to convert expressions to Polish suffix form. Then, defining the meaning of an expression to be its Polish suffix equivalent is the same as saying that we consider the meaning of an expression to be the algorithm it specifies for the computation of some function. Thus the *algorithmic* meaning of '(A)+(B)' is the same as that of 'A+B' and '(A+B)', since each specifies the same algorithm for computing the function $f(A, B) = A + B$; whereas 'A+(B+C)' and '(A+B)+C' have different algorithmic meanings, since each specifies a different algorithm for the computation of $f(A, B, C) = A + B + C$; of course, 'A+B' and 'A*B' have different algorithmic meanings, since these expressions represent different functions.

†In this chapter we have not introduced the concepts and notation necessary to make this distinction rigorous by means of *free* and *bound* variables.

As can be seen from this discussion, there are several possible definitions we might give for the semantics of arithmetic expressions, and the particular one to be chosen depends on the use of that expression. The literal semantics would be useful if we consider an expression only as a sequence of alphanumeric characters; the algorithmic semantics would be appropriate if we were compiling arithmetic expressions; and the functional semantics would be required if we were analyzing the expressions as mappings from one set into another. Yet, even though each of these semantics is useful under certain circumstances, none captures the meaning "usually" ascribed to arithmetic expressions. The problem is expressions such as

$$\text{'A}*(A-1)/(A-1)\text{'}.$$

This expression would usually be considered semantically equivalent to the expression

$$\text{'A'}.$$

Clearly, it is not equivalent under the literal or algorithmic semantics; but neither is it equivalent under the functional semantics, since

$$f(A) = \frac{A(A-1)}{A-1}$$

is not defined for $A = 1$, while $g(A) = A$ is. Thus we are led to a fourth definition of semantics of arithmetic expressions, *pragmatic semantics*. The functions in this example are not the same unless we define $f(1) = 1$; since we would normally make such a definition, we want these expressions to be considered equivalent. Thus pragmatic semantics is the same as functional semantics, with the added proviso that if E is some nonconstant arithmetic expression, then E/E has the same meaning as '1'.

In manipulating arithmetic expressions with a computer, we frequently want the computer to determine whether or not two expressions are equivalent; since each definition of semantics of expressions induces a different definition of "equivalent," we must examine each semantics individually. The problem of deciding equivalence is trivial when we use literal semantics: two expressions are equivalent if and only if they are identical, character for character. For algorithmic semantics, the same problem is not too difficult: translate each expression into Polish notation, and compare these Polish strings character for character. The determination of functional or pragmatic equivalence, however, is fairly difficult. Perhaps the only way to test equivalence in these cases is to establish some sort of *normal form* for expressions so that while each function has infinitely many expressions which represent it, only one of those expressions is in normal form. For example, for our arithmetic expressions we could specify that an expression is in normal form if it looks like P/Q, where P and Q are both *polynomial* functions in which the order of the terms is specified. We are then left with the task of translating expressions into the equivalent normal form and comparing them character by character.

The next section introduces some transformations that might be part of a collection of transformations for computing the normal form of an expression; these same transformations are, in some sense, simplifying in that the expression usually has a simpler form after applying a transformation. It is interesting to note that the concept of normal form cannot always be used in richer classes of expressions. For example, if we allow not only rational functions as expressions, but also allow the absolute value function, the transcendental constant π, and the transcendental function $\sin x$, then it can be rigorously proved, in the sense of Chapter 6, that no algorithm exists to decide the (functional) equivalence of such expressions; in this case, even if one could define a meaningful normal form, it would be ineffective.

1.4. SIMPLIFICATION

Arithmetic expressions such as we have described are associated with computers both as instructions to follow and as objects for manipulation. The former application came into existence with the first high-level programming languages such as FORTRAN. The latter application, particularly the large-scale development of programs for algebraic manipulation of expressions came somewhat later, but now occupies an important position among computer applications and is a current area of research. Here one finds the computer applied to tasks that are the less glamorous parts of mathematical problem solving, the pencil pushing; examples are multiplication of large polynomials, inversion of matrices with symbolic entries, simplification of expressions, symbolic differentiation and integration, and others. Of these, simplification of expressions is the most universal, and it can often spell the difference between a practical computation and an impossible one in terms of time and memory resources; the following example will illustrate these points.

In the beginning course of calculus one is introduced to differentiation through the concept of a limit applied to a certain difference quotient. What is usually retained from such a course is a collection of formal rules that are introduced for differentiating polynomial and rational expressions; with these rules the student is able to "mechanically" differentiate the expressions he encounters. In practice, one finds that differentiation, as a process, is less amenable to numerical methods than other processes, such as integration, for which there exist numerical techniques that apply to wide classes of functions, and where such methods are productive even when formal methods do not exist. Fortunately, however, formal differentiation can be applied to almost any expression, so that derivatives for most expressions can be obtained symbolically and then evaluated numerically in a straightforward way.

Here we present a procedure for differentiating an arithmetic expression in the parenthesis language of Section 1.1. Given such an expression and given a specified variable of that language, the procedure will produce a second arithmetic expression, the derivative of the given one with respect to the specified variable. Thus for each variable, say X, we define a function on the set of arithmetic expressions whose value on an arithmetic expression S is the derivative of S with respect to X, denoted

$$D\langle X, S \rangle.$$

For example, suppose that the character string 'VAR' is a variable in the language. Then the string 'VAR$*$Y$+$3' is an arithmetic expression and

$$D\langle \text{'VAR'}, \text{'VAR}*\text{Y}+3\text{'} \rangle$$

is the name in our description language for the arithmetic expression

$$\text{'}(1*\text{Y}+0*\text{VAR})+0\text{'}.$$

The following is the recursive definition of $D\langle X, S \rangle$ for any variable X and any arithmetic expression S of the language; there are seven cases:

If S is a constant, then '0'.

If S is a variable different from X, then '0'.

If S is X, then '1'.

If S has the form $S_1 S_2 S_3$, where S_1 is '(', S_3 is ')', and S_2 is an expression, then '(' $D\langle X, S_2 \rangle$ ')'.

If S has the form $S_1 S_2 S_3$, where S_1 is an expression, S_2 is an A-operator, and S_3 is a term, then $D\langle X, S_1 \rangle S_2 D\langle X, S_3 \rangle$.

If S has the form $S_1 S_2 S_3$, where S_1 is a term, S_2 is '$*$', and S_3 is a factor, then '(' $D\langle X, S_1 \rangle$ '$*$' S_3 '$+$' S_1 '$*$' $D\langle X, S_3 \rangle$ ')'.

If S has the form $S_1 S_2 S_3$, where S_1 is a term, S_2 is '/', and S_3 is a factor, then '(($' D\langle X, S_1 \rangle$ '$*$' S_3 '$-$' S_1 '$*$' $D\langle X, S_3 \rangle$ ')/(' S_3 '$*$' S_3 '))'.

Notice that if none of these cases applies, S is not an arithmetic expression. Also, note that in the last two cases new parentheses must be added around the expression to ensure proper interpretation in succeeding steps.

If we consider the string '$(R+H)/2*3.14$' discussed in Section 1.1, the derivation of $D\langle$'H', '$(R+H)/2*3.14$'$\rangle$ proceeds as follows:

$D\langle$'H', '$(R+H)/2*3.14$'$\rangle$

$= $ '$('D\langle$'H', '$(R+H)/2$'$\rangle$ '$*3.14+(R+H)/2*$' $D\langle$'H', '3.14'$\rangle$ '$)$'

$= $ '$((('$ $D\langle$'H', '$(R+H)$'$\rangle$ '$*2-(R+H)*$' $D\langle$'H', '2'$\rangle$ '$)/(2*2))*3.14+$ $(R+H)/2*0)$'

$= $ '$(((('$ $D\langle$'H', '$R+H$'$\rangle$'$)*2-(R+H)*0)/(2*2))*3.14+(R+H)/2*0)$'

$= $ '$((((('D\langle$'H', 'R'$\rangle$ '$+$' $D\langle$'H', 'H'$\rangle$ '$)*2-(R+H)*0)/(2*2))*3.14+$ $(R+H)/2*0)$'

$= $ '$(((((0+1)*2-(R+H)*0)/(2*2))*3.14+(R+H)/2*0)$'

We see that an algorithm for formally differentiating arithmetic expressions is quite easily described. The result, of course, although correct, is not in the form it would take if a person had done the calculation by hand; the obvious zero terms, unity multiples, and redundant parentheses would be eliminated in the early steps almost by second nature, resulting in a much simplified result:

$$\text{'}2/(2*2)*3.14\text{'},$$

or, even simpler,

$$\text{'}3.14/2\text{'}.$$

Apart from the obvious difference in intelligibility between the simplified and unsimplified derivatives, the simplification would have even greater impact in a program that required repeated differentiation. The reader who doubts this should apply the algorithm again to the unsimplified result. It is not unusual in practice for algebraic manipulation problems to involve expressions several pages in length; the size of the intermediate results in a computer program without some simplification built in would have a dramatic limiting effect on the size of the largest problem it could handle. A reasonable gauge of the worth of a symbolic manipulation algorithm to a computer user could be the extent to which the results are given in an easily understood, simple form.

Even though a person makes many simplifications almost automatically in the process of making a formal computation, simplification is by no means an easy task, as anybody who works with lengthy arithmetic expressions can verify. It is even more difficult to write down in the precise form of an algorithm what rules one is following when simplifying an expression. Yet this is precisely what must be spelled out explicitly if one wants a computer to achieve like results. The remainder of this section is devoted to a presentation of some rules for simplification of arithmetic expressions, which capitalize on the syntactical structure we have given them.

We shall use the parsing trees described earlier as a convenient description of the syntactical structure of arithmetic expressions. To simplify the figures, the following abbreviations will be used to label nodes of the trees: *ae* for ⟨arithmetic expression⟩, *t* for ⟨term⟩, *f* for ⟨factor⟩, *c* for ⟨constant⟩, and so on. A small triangle, $\triangle$, which may or may not be identified by a Roman numeral, will be used to indicate the presence of a nonempty subtree in a parsing tree where the details of the structure of the subtree are unimportant to the figure.

Perhaps the easiest and least controversial simplification problem is the removal of redundant parentheses. Consider the following pair of expressions:

$$\text{`(A+B)*(X+Y)',} \qquad \text{`(((A+(B)))*(((X+Y))))'.}$$

By counting parentheses we are able to tell that these two expressions both call for the same thing: the product of the sum of *A* and *B* and the sum of *X* and *Y*. Using the algorithm of Section 1.3 for translating parenthesis expressions into Polish suffix form, we find that they both translate into

$$\text{`A\_B\_+\_X\_Y\_+\_*'}$$

and hence have exactly the same algorithmic meaning. Clearly, the first expression is the simpler of the two, so any simplification algorithm should have the ability to reduce the second one to the first by removing the redundant parentheses. The two pairs of parentheses in the first expression are not redundant because the removal of either one would produce an expression with a different algorithmic meaning; for example, the expression

$$\text{`A+B*(X+Y)'}$$

translates into the Polish expression

$$\text{`A\_B\_X\_Y\_+\_*\_+'.}$$

How does one decide which pairs of parentheses are redundant and which are not, and how can this decision be made by a computer?

As was mentioned in Section 1.1, the productions we have used to define the arithmetic expressions in parenthesis form have the property that the syntactical structure of an expression with respect to these productions describes some of the semantics as well; in fact, it is sufficient to derive the algorithmic meaning of the expression. Thus we should be able to decide the question of redundant parentheses by looking at the syntactical structure,

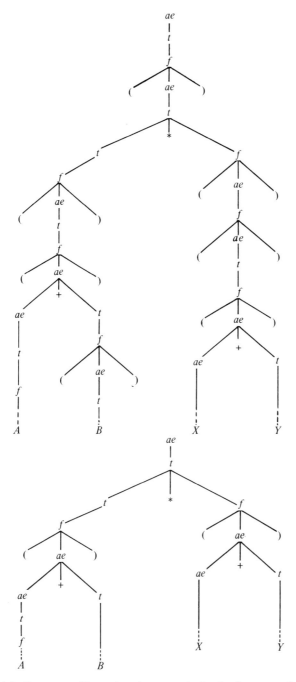

Fig. 1.5 Parse trees illustrating the removal of redundant parentheses.

the parsing tree. Figure 1.5 shows the parsing tree for the two expressions above. We see the basic structure of the expressions as the product of a term and a factor and the essential roles the parentheses play in forming this structure. It is required that the character strings 'A+B' and 'X+Y' be treated as a single term and a single factor, respectively, in the expression; this is impossible without parentheses, since 'A+B' and 'X+Y' by themselves have a two term structure, but it can be guaranteed by two pairs of parentheses, as the smaller tree illustrates. Once these parentheses are specified, any others are redundant, and in general we can say that a pair of parentheses is essential if it is needed to change the parse of a character string from ⟨arithmetic expression⟩ to ⟨term⟩ or ⟨factor⟩, or from ⟨term⟩ to ⟨factor⟩; otherwise, it is redundant. Considered in this light, redundant parentheses are easily identified in the parsing tree and may be removed using the following rule:

(1) If in the parsing tree there exists a subtree of the form

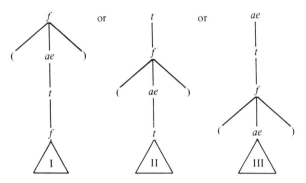

then the expression may be simplified by replacing that subtree with

respectively. The existence of such a subtree in a parsing tree indicates that the parentheses associated with the structure

do not ultimately change the parse of the character string they surround and hence are redundant and may be removed. Referring again to Figure 1.5, this rule calls for a total of five simplifications in the larger tree; the outermost parentheses and those around the 'B' are redundant, and there are an extra pair in the left main branch of the tree and two extra pairs in the right main branch.

The removal of redundant parentheses according to the preceding rule is clearly an equivalence-preserving transformation in the sense that the algorithmic meaning of an arithmetic expression does not change with its application. The reader is urged to check this by referring to Section 1.3 and the translation from parenthesis to Polish notation; see also Exercise 15.

While the parenthesis rule is neat and easy to apply, it would have only a slight effect on most arithmetic expressions if it were used alone; for example, in the expression

$$\text{`}((((0+1)*2-0*(R+H))/(2*2))*3.14+0*(R+H)/2)\text{'}$$

the outermost parentheses are the only ones that would be removed as redundant. The important simplification processes in the reduction of this expression to

$$\text{`}2/(2*2)*3.14\text{'}$$

are the elimination of zero terms and unity multiples. These also are simplifications that are done almost automatically by a person working with pencil and paper, and should be within the capabilities of any practical simplification algorithm. We present rules that reflect the properties of the zero expression, that is, the identity element property for addition and subtraction and dominance with respect to multiplication and division; the analogous rules for considering multiplications and divisions by unity are left as exercises for the reader.

The following transformations may be made in a parsing tree:

(2)   *may be replaced by*

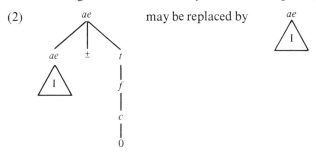

(3)   *may be replaced by*

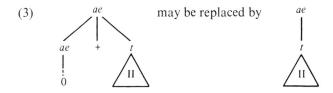

(4) 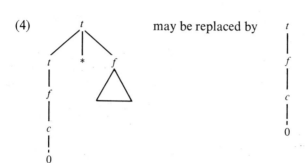 may be replaced by

(5) 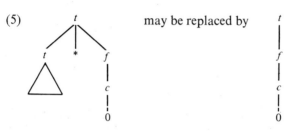 may be replaced by

The behavior of the nonzero expression with regard to division is considered in the following rule.

(6) If in a parsing tree there exists a subtree of the form

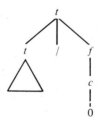

then the expression parsed by the tree cannot be evaluated; signal an error. Otherwise, any subtree of the form

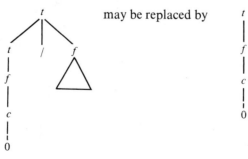 may be replaced by

The simplification of $D\langle$'H', '(R+H)/2*3.14'$\rangle$ using the above rules might proceed as follows:

Step	Rule	
		'(((($0+1$)*2-0*(R+H))/(2*2))*3.14$+0$*(R+H)/2)'
1	3	
		'((((1)*2-0*(R+H))/(2*2))*3.14$+0$*(R+H)/2)'
2	4	
		'((((1)*2$-$ 0)/(2*2))*3.14$+0$*(R+H)/2)'
3	4	
		'((((1)*2$-$ 0)/(2*2))*3.14$+$ 0 /2)'
4	6	
		'((((1)*2$-$ 0)/(2*2))*3.14$+$ 0)'
5	1	
		'(((1 *2$-$ 0)/(2*2))*3.14$+$ 0)'
6	2	
		'(((1 *2)/(2*2))*3.14$+$ 0)'
7	Exercise 16	
		'(((2)/(2*2))*3.14$+$ 0)'
8	1	
		'((2 /(2*2))*3.14$+$ 0)'
9	1	
		'(2 /(2*2) *3.14$+$ 0)'
10	2	
		'(2 /(2*2) *3.14)'
11	1	
		' 2 /(2*2) *3.14 '

or, '2/(2*2) *3.14'

As an interesting exercise the reader should follow this simplification through using the parsing trees to see how each rule applies.

    In contrast to the removal of redundant parentheses, the transformations just defined do not preserve the algorithmic meaning of the expression; for example, the algorithm specified by

$$\text{'A*1}+(1+0)\text{*B'}$$

is clearly not the same as that specified by the simplified version,

$$\text{'A+B'.}$$

However, these transformations are still equivalence preserving in that the same function is defined by an expression before and after any simplification using them. In the example, both expressions denote the same function of two variables, the summing function.

    The fact that rules (2) through (6) do preserve functional equivalence is, of course, a consequence of the properties of the zero element and the unit element of the field from which expressions can take their values, which we may take to be the field of real numbers. The binary operations of the field, addition and multiplication, satisfy all the field axioms, many of which induce functional equivalence-preserving transformations on arithmetic expressions in a natural way. The following rules for transforming parsing trees are of this type.

The commutative laws for addition and multiplication in a field are stated: for any two elements of the field, a, b,

$$a + b = b + a$$

and

$$a*b = b*a.$$

These two laws induce the following rules for transformations of a parsing tree.

In a parsing tree any subtree of the form

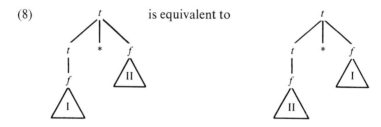

(7) ae is equivalent to ae

and either may replace the other.

(8) t is equivalent to t

and either may replace the other.

A close look will reveal that the tree transformations are slightly more restrictive than the axioms themselves would suggest. For example, the arithmetic expression

'A+B+C'

does not transform to

'C+A+B'

in one step as one might expect. However, by introducing a pair of redundant parentheses [applying rule (1) in reverse] one can get the expression

'C+(A+B)'

in two steps, and then remove the parentheses again by another rule.

The associative laws for addition and multiplication in a field are stated: for any three elements of the field, a, b, c,

$$a + (b + c) = (a + b) + c$$

and

$$a * (b * c) = (a * b) * c.$$

The corresponding tree transformation rules are

(9)   is equivalent to

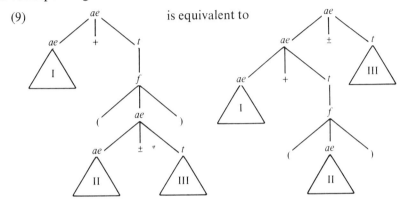

and either may replace the other.

(10)   is equivalent to

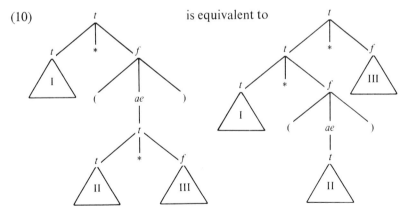

and either may replace the other.

These rules do not appear to effect simplification of an expression, but by repeated application they can aid the removal of parentheses that are not redundant by reducing the number of terms or factors within a pair of parentheses until the parentheses are redundant and removable by rule (1). For example, in the expression

'C+(A+B)'

the application of rule (9) yields

$$\text{`C+(A)+B',}$$

and rule (1) yields

$$\text{`C+A+B'.}$$

In a field multiplication is distributive over addition, satisfying the following identity: for any three elements of the field, $a, b, c,$

$$a*(b + c) = a*b + a*c.$$

As one might expect, the tree transformation rule corresponding to this axiom is slightly more complex than the previous ones.

(11)   is equivalent to

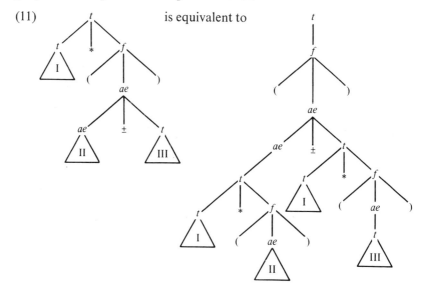

and either may replace the other.

The use of this transformation in a simplification algorithm is a matter of some controversy, largely because the question of which expression is simpler depends on what the expressions are being used for, that is, the context of the problem. For example, which of the two expressions

$$\text{`A*(1/B+1/C)'}\qquad\text{and}\qquad\text{`A/B+A/C'}$$

should be the final result of a simplification? If the next operation on this expression were to be a division by 'A', the first might be preferred, whereas if 'A/C' were to be subtracted next, the second would be more convenient.

The following rules reflect properties of the subtraction operator:

(12)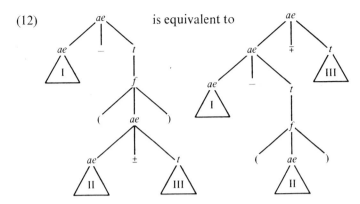

is equivalent to

and either may replace the other.

Note the change from $\pm$ on the left to $\mp$ on the right.

In a parsing tree any subtree of the form

(13)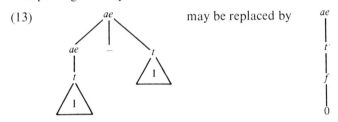

may be replaced by

The following example illustrates the interdependence of these rules and their use to simplify the expression 'A+B−A'.

Step	Rule	
		'A+B−A'
1	7	
		'B+A−A'
2	1	
		'B+(A)−A'
3	9	
		'B+(A−A)'
4	13	
		'B+(0)'
5	1	
		'B+0'
6	2	
		'B'

Several identities that yield several convenient transformation rules may be derived from the field axioms. They are sometimes considered as defining the arithmetic combinations of fractions, and induce the following rules:

(14)

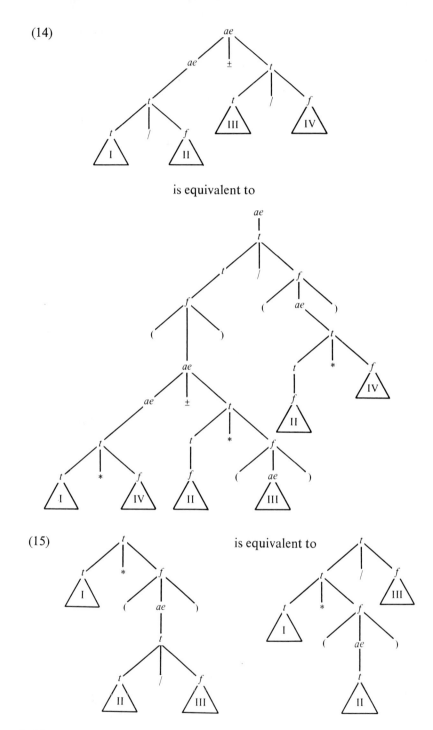

is equivalent to

(15)   is equivalent to

and either may replace the other.

34

(16)

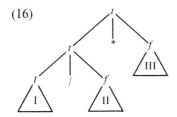 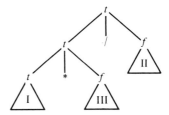

and either may replace the other.

(17)

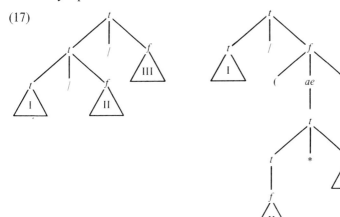

and either may replace the other.

All the transformations of arithmetic expressions defined so far preserve functional equivalence; the same function is defined by the expression before and after the transformation is applied. As mentioned in Section 1.3, there may be simplifications one would like to make that do not necessarily preserve this equivalence. An example could be the expression

$$\text{`X} * \text{X} * (\text{X} - 2)/(\text{X} - 2)\text{'},$$

which for all practical purposes could be considered equivalent to

$$\text{`X} * \text{X'},$$

the string '$(X-2)/(X-2)$' being considered as unity. Of course, this is not strictly true; nevertheless, factoring out common terms is a desirable capability of a simplification algorithm. The following rules will not necessarily preserve functional equivalence, but can be applied to effect pragmatic simplification.

(18) If the subtree f is not equivalent to

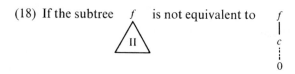

then 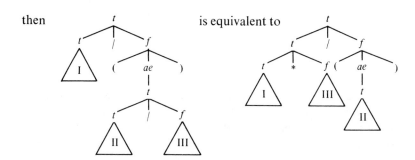 may be replaced by

(19) If the subtree f is not equivalent to ·

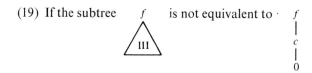

then   is equivalent to

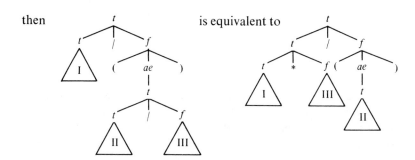

Figure 1.6 shows the reduction of

$$\text{'}A/(B/C)/(D/(E/F))\text{'}$$

to the form P/Q, where P and Q have no division signs, using the above rules.

1.5. REMARKS AND REFERENCES

The technique of defining the syntax of arithmetic expressions by means of productions is a slightly modified version of Backus–Naur form (BNF).

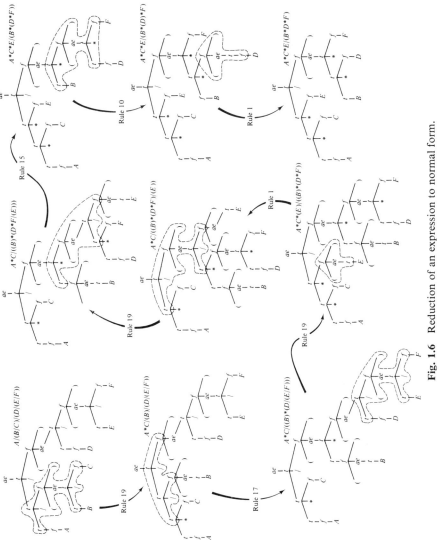

Fig. 1.6 Reduction of an expression to normal form.

37

This method was first used to define the programming language Algol 60 in

> NAUR, P. (ed.). "Report on the Algorithmic Language Algol 60,"
> *Comm. ACM, 3* (1960), 209–314.

It has now become a fairly standard technique for the definition of certain parts of the syntax of new programming languages. The difficulties of such a definition of a full programming language, as opposed to a small subset such as the one chosen in this chapter, can be appreciated by following the continuing development of Algol. See

> NAUR, P. (ed.). "Revised Report on the Algorithmic Language
> Algol 60," *Comm. ACM, 6* (1963), 1–17.

> KNUTH, D. E. "The Remaining Trouble Spots in Algol 60," *Comm.*
> *ACM, 10* (1967), 611–618.

Backus–Naur form is also substantially the same as the concept of context-free grammar introduced by linguists at about the same time. For a survey of this field, see

> CHOMSKY, N. "Formal Properties of Grammars," in *Handbook of*
> *Mathematical Psychology*, Vol. 2, pp. 323–418, R. D. Luce, R. R.
> Bush, and E. Galanter (eds.), Wiley, New York, 1963.

> HOPCROFT, J. E., and J. D. ULLMAN. *Formal Languages and Their*
> *Relation to Automata*, Addision-Wesley, Reading, Mass., 1960.

A good discussion of the relationship between compiler writing, Polish suffix notation, and semantics can be found in

> GRIES, D. *Compiler Construction for Digital Computers*, Chaps. 11
> and 12, Wiley, New York, 1971.

Similar material can also be found in

> WEGNER, P. *Programming Languages, Information Structures, and*
> *Machine Organization*, Sec. 4.3, McGraw-Hill, New York, 1968.

Using computers to operate on mathematical formulas in a symbolic way, as opposed to a numerical way, is an area of great current activity; a recent survey that contains an extensive bibliography is

> MOSES, J. "Algebraic Simplification: A Guide for the Perplexed,"
> *Comm. ACM, 14* (1971), 527–537.

Other papers in that same issue are also of interest. Another survey and extensive bibliography, although not as recent, is

> SAMMET, J. E. "Formula Manipulation by Computer," in *Advances*
> *in Computers*, Vol. 8, pp. 47–102, F. L. Alt (ed.), Academic
> Press, New York, 1967.

Among current programming languages, SNOBOL is the one in which it is easiest to write syntax-directed algorithms of the kind we have described in Section 1.4. See

> GRISWOLD, R. E., J. F. POAGE, and I. P. POLONSKY. *The SNOBOL4 Programming Language*, 2nd ed., Prentice-Hall, Englewood Cliffs, N.J., 1971.

1.6. EXERCISES

1. Construct the parse trees for the following arithmetic expressions:
   (a) X∗(Y+2.0)
   (b) ((BETA/0.5)∗C∗C)
   (c) X∗X+3∗X+4
   (d) A/B/C∗D
   (e) 6.2∗PI−(M−3)/0.3

2. Design an algorithm to determine whether an arbitrary finite character string in the alphabet of Section 1.1 is a legal arithmetic expression as defined by the productions of Section 1.1.

3. If we consider the set of arithmetic expressions defined in Section 1.1 only as strings of characters with certain syntactical properties, ignoring such semantic considerations as the precedence of operators, they can be generated by fewer productions. Design such a set of productions.

4. Modify the productions of Section 1.1 for arithmetic expressions as follows:
   (a) To allow '−' to be used as a unary operator denoting "additive inverse"; that is, extend the language to allow expressions such as

   $$-3.5, \qquad A+(-B), \qquad -X/Y,$$

   but not to allow the occurrence of consecutive operators; that is, rule out expressions such as

   $$A+-B, \qquad X/-Y, \qquad ---4.$$

   (b) To allow '↑' to be used as a binary operator denoting exponentiation; for example, $3x^2 + y^2$ would be written as 3∗X↑2+Y↑2. Your productions should have X↑Y↑Z parse as X↑(Y↑Z).

5. Prove that to each arithmetic expression, S, as defined by the productions in Section 1.1, there corresponds exactly one derivation tree.

6. The uniqueness of the derivation tree for each arithmetic expression in the parenthesis notation of Section 1.1 is a consequence of the set of productions used to define the language; such a set of productions is called *unambiguous*. On the other hand, if a language defined by a set of productions contains an expression that has more than one possible derivation tree, the set of productions is called *ambiguous*. Find an ambiguous set of productions that defines the same set of arithmetic expressions as the productions of Section 1.1, and give an example of an arithmetic expression that has more than one derivation tree.

7. Convert the expressions in Exercise 1 into Polish suffix notation and construct their parse trees.

8. Let $s_1 s_2 s_3 \ldots s_n$ be a sequence of characters in which each s_i is either an operand or a binary operator $+$, $-$, $*$, or $/$. Let c be a function defined by

$$c(0) = 0$$

$$c(i) = \begin{cases} c(i-1) + 1 & \text{if } s_i \text{ is an operand} \\ c(i-1) - 1 & \text{if } s_i \text{ is an operator} \end{cases}$$

For example,

	s_1	s_2	s_3	s_4	s_5	s_6	s_7
	X	Y	Z	$+$	W	$+$	$*$
0	1	2	3	2	3	2	1
$c(0)$	$c(1)$	$c(2)$	$c(3)$	$c(4)$	$c(5)$	$c(6)$	$c(7)$

Prove that a sequence $s_1 s_2 \ldots s_n$ is a legal Polish suffix expression if and only if

$$c(i) \geq 1 \quad \text{for } i = 1, \ldots, n$$
$$c(n) = 1$$

9. Design a set of productions similar to those in Section 1.2 that describe Polish prefix expressions, in which the operator precedes the operands.

10. State and prove an analog of the theorem in Exercise 8 for Polish prefix expressions.

11. Find a relationship between the Polish suffix and prefix forms of an expression. Use this relationship to design an algorithm to convert normal parenthesis notation into Polish prefix notation. (*Hint:* Consider translating the infix expression to suffix while reading from right to left instead of left to right.)

12. Design algorithms to evaluate Polish prefix and normal parenthesis expressions.

13. Consider the following alternative definition of $D\langle X, S \rangle$:

    If S is a constant, then '0'.

    If S is a variable different from X, then '0'.

    If S is X, then '1'.

    If S has the form '(S_1)', then '($D\langle X, S_1 \rangle$)'.

    If S has the form $S_1 S_2 S_3$, where S_1 is an expression, S_2 is an A-operator, and S_3 is a term, then $D\langle X, S_1 \rangle S_2 D\langle X, S_3 \rangle$.

    If S has the form $S_1 S_2 S_3$, where S_1 is a term, S_2 is '$*$', and S_3 is a factor, then

$$\text{'(} D\langle X, S_1 \rangle \text{ '}*\text{' } S_3 \text{) } + \text{ (} S_1 \text{ '}*\text{' } D\langle X, S_3 \rangle \text{)'.}$$

If S has the form $S_1 S_2 S_3$, where S_1 is a term, S_2 is '/', and S_3 is a factor, then

$$ \text{'((}D\langle X, S_1\rangle\text{ '*' }S_3\text{ '}-\text{' }S_1\text{ '*' }D\langle X, S_3\rangle\text{ ')/(' }S_3\text{ '*' }S_3\text{ '))'.} $$

If none of these cases apply, S is not an arithmetic expression.

(a) Show by an example that these rules are faulty; that is, give an arithmetic expression for which this definition gives an erroneous derivative.

(b) Modify these rules in at least three different ways so that they are correct, and prove the correctness of each of your modified set of rules.

14. (a) Design a set of productions to generate the set of polynomials in the variable X with integer coefficients and nonnegative integer exponents; for example

$$ X, \qquad -5, \qquad 3*X{\uparrow}2-2, \qquad X{\uparrow}5-X+X{\uparrow}11 $$

are valid polynomials.

(b) Describe an algorithm like the differentiation algorithm in Section 1.4 which integrates symbolically with respect to X any polynomial of the kind generated by your productions of (a).

(c) Assume that you would like to define an algorithm to symbolically integrate with respect to X any arithmetic expression in the language of Section 1.1. Discuss the following questions: Could it be done? How would you approach the problem? What difficulties would you anticipate?

15. Prove that repeated application of transformation (1) in Section 1.4 preserves algorithmic semantics.

16. Complete the simplification rules of Section 1.4 by defining rules that reflect the multiplicative identity property of the number 1.

17. The algorithms of Section 1.4 are all formulated in terms of the syntax of the arithmetic expressions, so all the rules are described in terms of the parsing trees for expressions. To permit implementation of these algorithms on a computer, it would be necessary to have some way of deriving the syntactical structure of an arbitrary arithmetic expression, which is determined by the operators in the expression. In particular, the operators occur in the derivation tree from the top down in roughly the reverse of the order they would be carried out in an evaluation of the expression. The operator that denotes the final operation in the evaluation is at the highest level in the tree, and thus is the *key* to the syntactical structure of the expression. Any derivation of the syntactical structure of an expression must identify this operator as a step in the derivation. In Polish suffix notation this is accomplished easily, since this is precisely the rightmost operator in the expression.

(a) Describe an algorithm that finds this key operator in an arithmetic expression in the parenthesis notation of Section 1.1.

(b) Describe a recursive procedure that will derive the syntactical structure of an arbitrary arithmetic expression in parenthesis notation based on an identification of the key operator in an expression.

18. State simplification rules parallel to those in Section 1.4 for the language of Polish notation expressions. These should be considerably easier to state than the rules for the arithmetic expression language of Section 1.1. Why?

2 COMBINATORIAL COMPUTING

Combinatorial mathematics is primarily concerned with problems about finite sets. The field was given its name by Leibnitz in 1666 in his *Dissertatio de Arte Combinatoria*, although many of its problems and results are considerably older. It has recently enjoyed a renaissance with the use of high-speed computation as a research tool.

Many combinatorial problems can always be thought of as searching for elements that satisfy certain conditions in a finite collection; as such, they always have a straightforward (brute-force) solution, that is, to look at every element in the finite set. Of course, "finite" does not necessarily mean "small," and it is characteristic of combinatorial problems that even those which are simply stated can involve searching extremely large sets. Indeed, the word "combinatorial" might well replace "astronomical" as an adjective meaning extremely large. For instance, if we consider that the most distant observed galaxies are about 10^{10} light-years away, a light-year is about 10^{16} meters, and a meter is about 10^{10} times the diameter of an atom, then we can estimate the ratio of the diameter of the observable universe to the diameter of an atom to be about 10^{36}. Although this number is very large, it is easy to find commonplace "combinatorial" numbers that are much larger; for example, the number of different possible pictures on a current television tube is between 10^{1000} and $10^{10,000}$.

The sheer size of these numbers means that in practice it is almost never possible to solve a combinatorial problem by brute force; one must be clever at exploiting every feature of the problem that will allow a shortcut. Some theoretical understanding of the problem (such as the use of symmetries to reduce the size of a search) is usually essential for a successful solution.

The importance of methods for dealing with combinatorial problems

43

has greatly increased in the last decade because of the widespread use of high-speed computers. One reason is that many important combinatorial problems could not be handled without computers, whereas with computers they have become feasible. Another reason is that when numerical problems (such as are discussed in Chapter 5) are put on a computer, they often end up becoming combinatorial problems.

A number of computing techniques applicable to a wide range of combinatorial problems are beginning to emerge, and some of these are discussed in this chapter. The backtrack algorithm of Section 2.1 is the most fundamental and most generally applicable; Sections 2.2 and 2.3 present some applications of backtrack. The theory of graphs is a subject in combinatorial mathematics experiencing an expanding range of applications; some of the specialized algorithms for dealing with graphs are discussed in Section 2.4. Section 2.5 is an introduction to sorting, a problem often considered to be trivial by the uninitiated, but one which is, instead, both mathematically profound and of great practical importance. If one looked at what all computers in the world were doing at any given moment, a fairly large fraction of them would be found sorting!

2.1. BACKTRACK

Combinatorial problems are characterized by the fact that they involve questions such as "how many?" or "list all such and such." Some familiar examples are the determination of the number of ways a quarter can be changed into pennies, nickels, and dimes, the number of paths along edges from one corner of a cube to the opposite corner, the number of solutions in positive integers of the equation $x_1 + x_2 + \cdots + x_m = n$, or the maximum number of queens which can be placed on a chessboard so that no queen controls a square occupied by another. Algorithmic methods for solving such combinatorial problems are fundamentally nothing more than procedures for generating the elements of a finite set, as defined by the problem specification, testing them according to the specified conditions, and displaying, listing, or counting those elements of the set found to satisfy the conditions.

For example, an inefficient but straightforward way to solve the change-making problem would be to generate all collections of dimes, nickels, and pennies containing at most 25 coins (this set is certain to contain all possible solutions to the problem) and count those whose value totals to 25 cents. This problem involves a set which is small enough so that it can be conveniently manipulated by hand: there are only 12 solutions.

On the other hand, the manual solution of the queens problem is not as feasible, but the basic approach would be the same. Recalling that in chess a queen has control over all squares in its row, column, and in the two diagonals intersecting at its position, it is clear that no more than eight queens can be placed safely on the board, for otherwise two queens would necessarily be in the same row; the problem is to discover whether it can be done with eight. As before, the most direct approach is to generate all possible configurations of eight queens on a chessboard and test for a solution among them, but here the number of configurations is very large (8^8, even when considering only those with exactly one queen in each column). We shall consider this problem in more detail later in this section, but first let us discuss, in general terms, the problem of searching through a large finite set.

Organizing a search or enumeration of a large finite set is not an easy problem. Usually, the search must be exhaustive, particularly in a problem such as the queens problem where there is some doubt about the existence of a solution. Moreover, we must avoid infinite loops in which the same elements are repeatedly examined. If the set possesses a natural linear ordering, it may be easy to implement an exhaustive search or enumeration by simply following that natural order. Thus searching the positive integers less than 1 million for those equal to the sum of squares of their digits is easily organized by testing the integers in order of increasing magnitude. In other cases, where a natural order is hard to discern or computationally difficult to follow, it may be easier and more efficient to organize a search by imposing an *artifical* linear order on the set.

Consider the following example. Let S be the set of all rational numbers that can be expressed as fractions n/d, in lowest common terms, where $n + d \leq 100$. Since the elements of S are all rational numbers, they are naturally ordered according to their magnitudes:

$$\frac{1}{99}, \frac{1}{98}, \frac{1}{97}, \ \cdots \ , \frac{1}{49}, \frac{2}{97}, \frac{1}{48}, \frac{2}{95}, \ \cdots \ , \frac{99}{1}.$$

From a computational point of view, however, this is an inefficient order in which to generate the set. The inefficiency can be appreciated by considering, for example, the difficulty of deciding what number should come after $\frac{2}{3}$. Let us instead define another order on this set of numbers by saying that

$$\frac{n_1}{d_1} \text{ precedes } \frac{n_2}{d_2} \text{ if and only if } \begin{cases} n_1 + d_1 < n_2 + d_2, \\ \qquad\quad \text{or} \\ n_1 + d_1 = n_2 + d_2 \text{ and } n_1 < n_2. \end{cases}$$

Here $<$ denotes the natural order (less than) on the positive integers. This ordering of S is

$$\frac{1}{1}, \frac{1}{2}, \frac{2}{1}, \frac{1}{3}, \frac{3}{1}, \frac{1}{4}, \frac{2}{3}, \ \cdots \ , \frac{97}{3}, \frac{99}{1}.$$

The following algorithm efficiently generates the elements of the set S in this order:

Step 1 (Initialize)
Set SUM←1.

Step 2 (Increment SUM)
Set SUM←SUM+1. If SUM > 100, we are done; otherwise, set N←0.

Step 3 (Increment numerator)
Set N←N+1; if N=SUM, go back to step 2.

Step 4 (Generate fraction)
Set D←SUM−N. If N and D are relatively prime, insert the fraction N/D in the sequence. In any case, return to step 3.

Frequently, the most efficient search in a combinatorial problem is obtained by structuring the elements of the underlying set as a tree, the same structure that we used in the previous chapter to study the syntax of arithmetic expressions. Figure 2.1 illustrates the organization of the set T of fractions

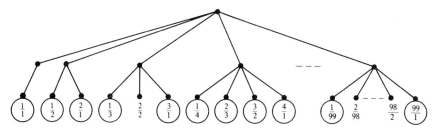

Fig. 2.1   Tree organization of a superset of S. The elements of S are circled.

n/d such that n and d are positive and $n + d \leq 100$, but which may not be in lowest common terms; clearly, S is a proper subset of T. Each element of the set S occurs exactly once as the label of a leaf in this tree. The enumeration of the elements of S by the previous algorithm is easily described in terms of the tree: starting at the root, follow the leftmost branches; when you come to a leaf, list the fraction found there, provided it is in lowest common terms, then back up the tree until you can again proceed downward along the leftmost branch, which was not previously followed.

This process is the essence of a technique called *backtrack* or *depth first search*, and also of a similar technique called *branch and bound*. It is a natural algorithmic tool which has been discovered and employed independently by many researchers before it was abstracted and formally described in the

1950s; it is regarded as one of the fundamental tools of combinatorial computing. We shall describe a backtrack algorithm in rather abstract terms; the reader should return to this concise and general statement of the algorithm later when he encounters specific applications of it.

Any combinatorial problem to which the backtrack algorithm can be applied fits the following general description: given n linearly ordered sets $U_1, U_2, \ldots, U_n$, we want to construct a vector $A = (a_1, a_2, \ldots, a_n)$, where $a_1 \in U_1, a_2 \in U_2, \ldots, a_n \in U_n$, which satisfies a given set of conditions or constraints.

In the backtrack algorithm, the vector A is built up one component at a time from left to right. Suppose that we already have assigned values to the first $k - 1$ components,

$$A = (a_1, a_2, \ldots, a_{k-1}, ?, \ldots, ?);$$

then the set of constraints will limit the choices for the next component a_k to some subset S_k of U_k. If S_k is nonempty, we are free to choose a_k to be the smallest element of S_k and then continue with S_{k+1}, and so on. If, however, the conditions imply that S_k is empty, then we backtrack to the previous stage, discard the $(k - 1)$st element a_{k-1} and choose as a new a_{k-1} the next element of S_{k-1} larger than the one we have just rejected. With this new choice of a_{k-1}, the conditions of the problem may now allow S_k to be nonempty, and so we try again to choose an a_k. If no choice for a_{k-1} is possible, we backtrack farther to get a new a_{k-2}, and so on. More explicitly, this algorithm is

Step 1 (Initialize)
Set $k \leftarrow 1$ and $S_1 \leftarrow U_1$.

Step 2 (Next element of S_k)
If S_k is empty, go to step 5. Otherwise, set a_k equal to the smallest element in S_k.

Step 3 (Solution complete?)
If the solution is complete (i.e., $k = n$), then record $(a_1, \ldots, a_n)$ as a solution. If all solutions are to be found, set $k \leftarrow k + 1$ and go to step 5; otherwise, stop.

Step 4 (Increment k)
Increment k by 1 and compute S_k, and go back to step 2.

Step 5 (Backtrack)
If $k = 1$, we can backtrack no farther, so stop, having found all solutions, or that none exists. If $k > 1$, decrement k by 1 and compute S_k. Remove all previously used elements of S_k by setting $S_k \leftarrow S_k - \{a \mid a \leq a_k\}$ and return to step 2.

Let us return to the chess-queens problem discussed earlier. We first consider a very natural manual procedure that could be applied to the problem and then see how it fits the abstract description of backtrack. Remember that no more than eight queens can possibly be placed in a configuration satisfying the conditions of the problem; thus the question is whether or not there is a solution with exactly eight queens. The manual search for such a solution could begin by placing the first queen in the lower-left-hand corner square of the board. Since we know that a solution with eight queens would necessarily place one queen in each column, we can insist that the second queen be placed in the second column, the third queen in the third column, and so on, moving from left to right. To impose an order on the search, we shall agree that in each column the first position to be tried will be the lowest one possible. Figure 2.2(a) illustrates the positions of the first five queens placed by this procedure. At this point all the squares in the sixth column are controlled by the five queens already on the board, so that it is necessary to backtrack to find a new partial configuration of these five before we shall be able to place a sixth queen on the board. Figures 2.2(b) and (c) illustrate the two necessary backtrack steps. The reader should complete this example on his own.

It is easy to see how this algorithm can be fitted to the formal description: For $i = 1$ to 8, let $U_i = \{1, 2, 3, 4, 5, 6, 7, 8\}$, the numbers being ordered by magnitude. U_i can be thought of as the row numbers on the eight by eight chessboard, numbered from bottom to top. We seek vectors $A = (a_1, a_2, a_3, a_4, a_5, a_6, a_7, a_8)$ such that if $i \neq j$, then

1. $a_i \neq a_j$.
2. $a_i - i \neq a_j - j$.
3. $a_i + i \neq a_j + j$.

For each i, a_i is the row number of the queen in the ith column, and the conditions are interpreted as

1. No two queens lie in the same row.
2. No two queens lie on the same ascending diagonal ($a_k = k +$ constant).
3. No two queens lie on the same descending diagonal ($a_k = -k +$ constant).

The computation of the sets S_k in steps 4 and 5 of the algorithm is done by first defining

$$S_1 = U_1 = \{1, 2, 3, 4, 5, 6, 7, 8\};$$

then if $a_1, a_2, \ldots, a_{k-1}$ have been chosen, S_k is obtained by removing all

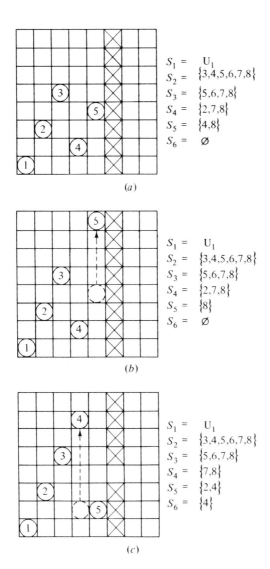

Fig. 2.2   Using the backtrack procedure to solve the queens problem.

elements a_i for $i < k$, all elements $a_i - i + k$ for $i < k$, and all elements $a_i + i - k$ for $i < k$. Figure 2.2 shows the various S_i constructed at the illustrated steps; notice that a_i is always the smallest element in S_i. (The computation of the sets S_k is reminiscent of another useful combinatorial method, *sieves*. This technique is related to backtrack in that whereas the backtrack procedure computes sets of solutions, a sieve computes sets of *nonsolutions*. Sieves are useful in certain number theoretic investigations and are discussed in that context in Section 5.3.1.)

Several applications of the general backtrack algorithm are found in the following sections and in the exercises. The reader is urged to analyze each one in the same manner as the previous example to appreciate the great flexibility of this important programming tool.

2.2.   BLOCK DESIGNS

A block design, in the most general meaning of the term, is simply an arrangement of objects into sets (called blocks), subject to certain conditions. With such a loose definition, one should expect to find the concept of a block design broadly interpreted and applied in many different areas. This is indeed the case, but one of the most important historical motivations for studying them, apart from the interesting questions they pose for combinatorial mathematics, has been their application to the design of statistical experiments; from this have come widely accepted "standard" conditions for defining certain types of block designs. In Section 2.2.1 we consider one of these types, the balanced incomplete block design. In Section 2.2.2 we discuss latin squares, an area where high-speed computers have recently become almost indispensable for research.

2.2.1.   Balanced Incomplete Block Designs and Statistical Experiments

Since about 1935, block designs have had an important application in the theory of experimental design, the study of the construction of experiments for making (usually comparative) evaluations of certain entities with a view to their statistical analysis. An experiment designed to compare, for example, the relative merits of several remedies for the common cold, or agricultural soil treatments, can be viewed as a block design, on which certain conditions have been imposed depending on the constraints represented by the physical situation in which the experiment is performed. One particular group of conditions that defines a type of design with relatively wide applicability yields *balanced incomplete block designs*, whose properties can be conveniently described in terms of some well-known concepts in mathematics.

Let us see how the conditions leading to a balanced incomplete block design arise naturally. Suppose that we wish to find out which one of a fairly large number (n) of competing products $A_1, A_2, \ldots, A_n$ will be preferred by consumers. Our technique will be to ask a number (b) of potential consumers $B_1, B_2, \ldots, B_b$ to examine these products and to report their findings by making judgments of the type "A_i is better than A_j" whenever they clearly prefer one product to another. Now there may arise certain constraints that will limit the ways in which the products should be distributed to the consumers. For example, if we are testing cold capsules, it would be unreasonable to ask a person to try out more than two or three different capsules in the course of one illness, and in light of the wide variations in the symptoms and severity from one cold to the next, it is also unreasonable to consider comparisons between the effects on different illnesses. We can reasonably assume that the number of products is too large to expect a consumer to examine each product.

Under these conditions we see that if we wish to carry out this experiment so as to avoid certain types of bias (e.g., if certain products or pairs of products were not examined as often as others), we shall insist on the following three conditions:

1. Each product must be examined by the same number (q) of consumers.
2. Each consumer must examine the same number (k) of products.
3. Each (unordered) pair of products must be examined by the same number (p) of consumers.

An arrangement of objects (products) into blocks (consumers) that satisfies these three conditions is a balanced incomplete block design; "incomplete" means that not every one of the $\binom{n}{k}$ sets of k objects out of n occurs as a block, and "balanced" refers to the fact that all pairs of objects must occur equally often.

A convenient way to visualize block designs is by means of an *incidence matrix*, M, whose rows correspond to objects and whose columns correspond to blocks; the element m_{ij} is 1 or 0, depending, respectively, on whether or not object A_i is in block B_j. The conditions above can then be restated as

1′. Every row must contain q ones.
2′. Every column must contain k ones.
3′. Each pair of rows must have exactly p ones in common.

For example, a block design with $n = 3$, $b = 6$, $k = 2$, $q = 4$, and $p = 2$ is

$$M = \begin{array}{c|cccccc} & B_1 & B_2 & B_3 & B_4 & B_5 & B_6 \\ \hline A_1 & 1 & 1 & 1 & 1 & 0 & 0 \\ A_2 & 1 & 1 & 0 & 0 & 1 & 1 \\ A_3 & 0 & 0 & 1 & 1 & 1 & 1. \end{array}$$

It is easy to see that the five parameters n, b, k, q, and p of a balanced incomplete block design must satisfy two equations:

$$bk = nq,$$
$$p(n - 1) = q(k - 1).$$

The first comes from counting the total number of 1's in an incidence matrix, the second from counting the number of pairs in which A_1 occurs: A_1 is paired with each of the other $n - 1$ objects exactly p times; also, A_1 occurs in q blocks, and in each of these is paired with $k - 1$ other objects. These two basic relations are necessary but *not* sufficient for the existence of a balanced incomplete block design with a given set of parameters.

Because of their direct applications to experimental design, balanced incomplete block designs have been extensively catalogued and methods have been developed for constructing many specific types. However, the general question of the existence of designs with given parameters remains open, since no method of solution is known other than an exhaustive search.

The computational approach to block designs is important, since in many applications one encounters problems of a combinatorial nature very similar to block designs, but which are not quite of the standard types that have been extensively studied. Since mathematical constructions in combinatorial theory tend to be applicable only to one type of problem, the methods developed for block designs may not work; on the other hand, computational techniques like the backtrack procedure are extremely general in their applicability. For example, if we wished to find a block design using the backtrack procedure, we could let the vector A be the set of all elements of the incidence matrix, ordered in some convenient way (by rows, say), and let $U_1 = U_2 = \cdots = U_n = \{0, 1\}$. Whether the constraints are that there are precisely q ones per row, or at most q ones per row, or a prime number of ones per row has little effect on the procedure. This approach is also important since there are many open problems in this area, some of which could be settled by exhibiting an explicit solution, or by finding a counterexample to a conjecture. As is the case elsewhere in combinatorial mathematics, the judicious use of a computer to aid in searching for such cases is almost a necessity.

2.2.2. Latin Squares and Scheduling Problems

Many problems concerned with scheduling events, constructing time-tables, assigning tasks or resources, and so on, are easily and naturally formulated in terms of block designs or similar arrangements. This is about all that can be said for real, practical problems, but if we idealize the problem to make it sufficiently regular (usually at the cost of realism!), it may be formulated in terms of latin squares. Latin squares have a long history in connection with mathematical puzzles, such as magic squares, and recently they have been the objects of considerable research with digital computers in connection with some open problems in the study of finite geometries.

A *latin square of order n* is an n by n matrix whose elements are chosen from the integers 1 to n in such a way that each such integer occurs exactly once in each column and once in each row. As an example of how a latin square might arise, consider a school with four teachers, T_1, T_2, T_3, and T_4, and four classes of students, C_1, C_2, C_3, and C_4, meeting for four class periods per day, at 1, 2, 3, and 4 o'clock, respectively. Assume that every teacher must meet with each class for one time period per day. In this situation a possible schedule, an assignment of teachers and classes to time periods, can be made as shown in the following matrix; the number k in row i and column j means that teacher T_i meets class C_j at time k.

$$
A = \begin{array}{c|cccc}
 & C_1 & C_2 & C_3 & C_4 \\
\hline
T_1 & 1 & 2 & 3 & 4 \\
T_2 & 2 & 3 & 4 & 1 \\
T_3 & 3 & 4 & 1 & 2 \\
T_4 & 4 & 1 & 2 & 3.
\end{array}
$$

Since no teacher can meet two classes simultaneously, no integer (hour) can occur twice in the same row; since no two teachers can teach the same class simultaneously, no integer can occur twice in the same column. Only four class times are available, so it is easy to see that a proper schedule must be a latin square.

Latin squares are easy to construct, but it is possible to pose difficult combinatorial problems by imposing certain additional constraints on a square or a set of squares. Suppose, for example, that one day four groups of visitors appear at the school, wishing to sit in on the classes. Each group wants to see each teacher and each class, and no classroom is large enough to hold two groups of visitors at once. Under these conditions it is clear that just to assign the visitor groups to teachers and classes (without regard to the meeting times) requires the construction of another latin square; this time an entry l in row i and column j means that group l sits in when teacher T_i meets class C_j.

$$
B = \begin{array}{c|cccc}
 & C_1 & C_2 & C_3 & C_4 \\
\hline
T_1 & 1 & 2 & 3 & 4 \\
T_2 & 3 & 4 & 1 & 2 \\
T_3 & 4 & 3 & 2 & 1 \\
T_4 & 2 & 1 & 4 & 3.
\end{array}
$$

Now the crucial question: Is this assignment realizable; that is, can it actually be carried out under the given conditions? We shall see that in order for this to be true the two latin squares we have constructed must together satisfy a very stringent requirement. Consider the event "T_i meets C_j." The first latin square tells us that this happens at time k, and the second one tells us that group l is present. Now if there were another event "T_r meets C_s," ($r \neq i$ or $s \neq j$) with the same pair of entries k, l in the two squares, this could mean that group l must be present at two distinct class meetings simultaneously, which is impossible. Hence for the assignment to be realizable no two of the ordered pairs (A_{ij}, B_{ij}), $1 \leq i, j \leq 4$, can the same. Since there are exactly 16 different ordered pairs of this kind, each ordered pair occurs precisely once as i and j range independently from 1 to 4. If the two squares satisfy this requirement, they are said to be *orthogonal*. In general, two latin squares of order n are orthogonal if no two of the n^2 pairs of corresponding components are the same. Looking at the two squares, A and B, constructed in this example, we see that they are not orthogonal; in fact, each of four groups of visitors is caught in the position of having to be in two places at once. Unfortunately, the original class schedule (square A) was a particularly poor one to choose for this purpose (see Exercise 4). Orthogonal latin squares of order 4 do exist; here are three squares, any two of which are orthogonal:

$$
C = \begin{matrix} 1 & 2 & 3 & 4 \\ 3 & 4 & 1 & 2 \\ 4 & 3 & 2 & 1 \\ 2 & 1 & 4 & 3, \end{matrix} \qquad
D = \begin{matrix} 1 & 2 & 3 & 4 \\ 4 & 3 & 2 & 1 \\ 2 & 1 & 4 & 3 \\ 3 & 4 & 1 & 2, \end{matrix} \qquad
E = \begin{matrix} 1 & 2 & 3 & 4 \\ 2 & 1 & 4 & 3 \\ 3 & 4 & 1 & 2 \\ 4 & 3 & 2 & 1. \end{matrix}
$$

Orthogonal latin squares lie beneath many mathematical puzzles. For example, a well-known problem in recreational mathematics is to arrange the numbers from 1 to n^2 in an n by n matrix so that the sum of the elements in any row or column is the same constant, S. Such an array is called a *magic square*. A magic square can be generated from a pair of orthogonal latin squares: Consider the two latin squares C and D, just given; define a new 4 by 4 array M given by the formula

$$M_{ij} = 4(C_{ij} - 1) + D_{ij}.$$

We see that this formula produces exactly the numbers from 1 to 16, given the 16 ordered pairs (C_{ij}, D_{ij}) for $1 \leq C_{ij}, D_{ij} \leq 4$ (view $C_{ij} - 1$ and $D_{ij} - 1$ as numerals base 4). Furthermore, the sums of the rows

$$
\begin{aligned}
M_{i1} + M_{i2} + M_{i3} + M_{i4} &= 4(C_{i1} + C_{i2} + C_{i3} + C_{i4} - 4) \\
&\quad + (D_{i1} + D_{i2} + D_{i3} + D_{i4})
\end{aligned}
$$

are all the same constant (34), since C and D are both latin squares. The magic square M thus generated is as follows; notice that since each of the two diagonals of C and D contain all the integers 1, 2, 3, and 4, the sums of the diagonals of M are also equal to 34.

$$M =$$

1	6	11	16
12	15	2	5
14	9	8	3
7	4	13	10

The formula for the 4 by 4 magic square can be generalized to permit the construction of similar squares from any pair of orthogonal latin squares; see Exercise 5.

The name "latin square" for the matrices we have described here is said to have been derived from the fact that Leonhard Euler, circa 1780, used Roman letters for the symbols in his squares. Stimulated by an old mathematical puzzle, he established the existence of orthogonal pairs of latin squares for all orders *not* of the form $4n + 2$, but was unable to find any such pairs of order 6 despite a lengthy search. Ultimately, he conjectured that no orthogonal pairs of latin squares existed for order 6 or for any order of the form $4n + 2$. Euler's conjecture stood for centuries; the case for order 6 was actually proved only early in this century by an exhaustive search of all possibilities, a monumental task without computers. When computers became available, they were put to work on the problem for order 10, and, in this case, the computer arrived on the scene too late, for the conjecture was finally *disproved* by a general construction (using no computers) in 1959, which yielded orthogonal latin squares for any order $4n + 2 \geq 10$. Subsequently, relatively fast (considering the size of the combinatorial problem) programs were written to find many such pairs for order 10.

A set of latin squares is called *mutually orthogonal* if every pair of squares in the set is orthogonal; the three squares C, D, and E, given previously, are mutually orthogonal. A set of $n - 1$ mutually orthogonal latin squares of order n is called a complete set; it can be easily shown that this is the maximum possible number of mutually orthogonal squares (see Exercise 6). The problem of finding complete sets of mutually orthogonal latin squares has become prominent, since it turns out that such a set corresponds to a type of finite geometry known as a finite affine plane. The smallest order for which the question of the existence of a complete set is still unknown is again order 10. As was the case when computers were first applied to Euler's conjecture, some evidence has been obtained for the nonexistence of a set of more than two mutually orthogonal squares of order 10, but a definitive search has not yet been made due to the computer time it would require.

2.3. TILED RECTANGLES AND ELECTRICAL
NETWORKS

We have characterized combinatorial problems as those which study configurations of finite sets, so it is not surprising that such problems arise naturally in situations in which a finite set of physical objects is involved. Typical examples might be packing luggage into a small automobile trunk, the familiar situation in which objects seem to defy repacking in the original container, or the problem of the bricklayer who must arrange his 2 by 4 by 8 inch bricks into a wall, a fireplace, or a monument. These examples suggest the more general problem of completely filling a given volume of a given plane area with a finite set of objects of one kind or another. Such problems are known as *packing problems*, or (in two dimensions) *tiling problems*, and have been the subjects of some interesting combinatorial studies.

In this section we are primarily concerned with one particular type of problem, that of tiling, or completely filling, the interior of a rectangle with nonoverlapping squares. Problems in this area have received considerable attention by combinatorial theorists and have been successfully studied using the computer. Our interest in these problems is largely due to the close connection between tiled rectangles and graphs (in the form of electrical networks), an important subject discussed further in Section 2.4.

The first question to be answered is under what conditions is it possible to tile a rectangle with a finite set of squares? It is obvious that any h by w rectangle where the ratio $h:w$ is a rational number p/q can be tiled using pq congruent squares whose sides have length $h/p = w/q$. Conversely, if an h by w rectangle can be tiled with a finite set of congruent squares of any size whatever, the ratio $h:w$ is necessarily rational. Thus the question of tiling an h by w rectangle with congruent squares is easily settled: it is possible if and only if h/w is rational. To treat the general problem of tiling with any finite set of squares, it is convenient to attack the problem by considering electrical network analysis; the appropriateness of this approach will soon be apparent.

Figure 2.3 shows simple examples of planar electrical networks; they are planar because they can be drawn on a plane surface without any crossing lines. This can be considered to be a wiring diagram in which the lines represent wires through which an electric current can pass. These wires are connected with one node at the top, called the source, and one node at the bottom, called the sink; current flows into the network at the source and leaves the network at the sink. With each branch in this network we associate a constant, which represents the resistance of that branch to the passage of current, and in practice this constant is specified by inserting a resistor in the branch. The current that passes through the network is distributed among the branches of the network on its way from the source to the sink, and so we can speak of individual currents in the branches. The total voltage across the network, the

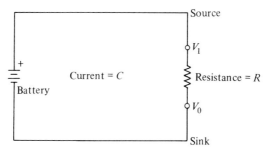

(a) The simplest possible network

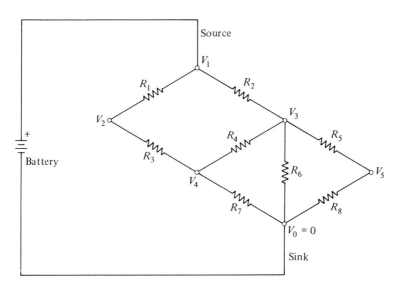

(b) A more complicated network

Fig. 2.3 Two planar electrical networks.

difference in voltage between the source and the sink, is also distributed in the sense that a voltage can be measured between each node and the sink, and this voltage depends on the position of the node in the network. The magnitudes of these individual currents and voltages are the object of network analysis.

Electrical currents in such a network are found to behave in accordance with several basic physical principles, two of which concern us here. The simplest possible network is shown in Figure 2.3(a); in this network we say that the current flows from the point of higher voltage V_1 to the point of lower voltage V_0, and the magnitude of this current (expressed in amperes) satisfies

Ohm's law:

$$\text{current from } V_1 \text{ to } V_0 = \frac{\text{drop in voltage from } V_1 \text{ to } V_0}{\text{resistance of branch}}$$

or
$$\text{current} = \frac{V_1 - V_0}{R},$$

where V_1 and V_0 are expressed in volts, and R in ohms. This law applies to any individual branch in an arbitrary network; in Figure 2.3(b) we have

$$\text{current in branch } R_5 = \frac{V_3 - V_5}{R_5}.$$

The second physical principle we need is known as Kirchhoff's first law, and it states that the algebraic sum of all currents into a given node is equal to zero; equivalently, the sum of the incoming currents at a given node equals the sum of the outgoing currents from the node. This means that there is no accumulation of electrical charge at a node.

Consider again the network of Figure 2.3(b). Suppose that we have fixed the value of each resistance at 1 ohm and the total current through the network is 1 ampere. Let us apply the two principles to find the total voltage drop across the network, $V_1 - V_0$. Notice first that by our choice of 1-ohm resistances, we have simplified the statement of Ohm's law to read

$$\text{current through a branch} = \text{voltage drop across branch}.$$

Without loss of generality we may let V_0, the voltage at the sink, equal zero. Since the total current flowing through the network flows through the sink we have

$$(V_3 - 0) + (V_4 - 0) + (V_5 - 0) = 1$$
or
$$V_3 + V_4 + V_5 = 1.$$

Similarly, for currents at the source node we have

$$(V_1 - V_2) + (V_1 - V_3) = 1$$
or
$$2V_1 - V_2 - V_3 = 1.$$

For the currents at the remaining nodes, we obtain

$$V_1 - 4V_3 + V_4 + V_5 = 0,$$
$$V_1 - 2V_2 + V_4 = 0,$$
$$V_2 + V_3 - 3V_4 = 0,$$
$$V_3 - 2V_5 = 0.$$

From the physical interpretation we expect that this system of equations has a unique solution. Furthermore, since the coefficients in the equations are all rational, the resulting solution will also be rational; in particular, the total voltage drop across the network has a rational value. Indeed, solving the system in this example yields

$$V_1 = \frac{31}{29}, \quad V_2 = \frac{21}{29}, \quad V_3 = \frac{12}{29}, \quad V_4 = \frac{11}{29}, \quad V_5 = \frac{6}{29}.$$

Consider now a slightly different representation of planar electrical networks. Redraw a planar network by representing each of its branches by a rectangle whose width w is the current flowing through the branch and whose height h is the voltage drop across the branch (assume the higher voltage to be at the top edge of the rectangle). Then, by Ohm's law, the resistance of a branch equals the ratio $h:w$ of the dimensions of the rectangle. In the example we have assumed that all resistances are 1 so that the rectangles representing them are actually squares. A node of the network becomes a rectangle of zero height, that is, a horizontal line segment whose length is the current flowing through the node. Each square is placed so that its top edge and bottom edge lie on the horizontal segments representing the nodes to which the corresponding branch is connected, as in Figure 2.4(b). We have now drawn a tiled rectangle whose width is the total current through the network and whose height is the voltage drop between the source and the sink of the network. In a completely straightforward manner we may reverse the correspondence and draw a planar electrical network of 1-ohm resistors corresponding to any rectangle tiled with squares.

This correspondence between tiled rectangles and electrical networks answers the original question about tiling an h by w rectangle with any finite set of squares: let such a tiling be given for an h by w rectangle; the network corresponding to the tiled rectangle can be analyzed exactly as before, solving a system of linear equations with integer coefficients. The solution gives a rational value for the total voltage drop across the network, in other words, a rational value for $h:w$ for the rectangle. Combining this result with the earlier one regarding tiling with congruent squares, we conclude that

an h by w rectangle can be tiled with a finite set of squares

if and only if

it can be tiled with a finite set of congruent squares

if and only if

the ratio of its height to its width is rational.

One of the more prominent tiling problems that has been considered is the categorization of the so-called perfect squares: a rectangle (square) is a

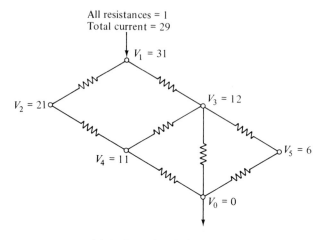

(a) A planar electrical network

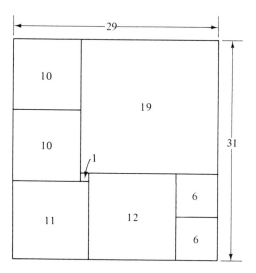

(b) A tiling of a rectangle which corresponds
to the network in (a)

Fig. 2.4 Transformation of a network problem into a tiling
problem.

perfect rectangle (square) if it is possible to tile it with a finite set of squares, no
two of which are congruent. Very large perfect squares have been discovered
in which many squares are used in the tiling, and some effort has been made
to find perfect squares that consist of a minimum number of smaller squares.
A computer search has established that no perfect squares exist using fewer

than 20 squares; perfect squares have been found that consist of 25 constitu-
ent squares, but to our knowledge none has been discovered using 24. Considerable interest has been drawn to this problem by the equations

$$1^2 + 2^2 + 3^2 + \cdots + 24^2 = 4900 = 70^2,$$

which make it conceivable that the 70 by 70 square might have a tiling using
exactly the squares of sizes 1 through 24. We do not know of a proof that this
is impossible, and the problem has been posed to maximize the area covered
in a 70 by 70 square by a partial tiling using only squares from the set of
squares of sizes 1 through 24. Trial-and-error methods have been used to
obtain results in connection with this question, and improving the known
results would probably require an exhaustive search of all possible partial
tilings using the squares allowed. Such a search would, of course, answer the
original question of the existence of a complete tiling, if it were computationally feasible.

    Let us see how a backtrack procedure could be employed in connection
with tiling rectangles. We shall consider the following question: Given a set
of squares whose combined area equals the area of a given rectangle, is there
an arrangement of these squares that will tile the rectangle? Note that a tiling
of a rectangle can be represented as a set of ordered pairs $((X, Y), Z)$, where
(X, Y) are coordinates of the lower-left-hand corner of a square and Z is the
length of one side of the square. Thus if we place the rectangle in a rectangular
coordinate system with its lower-left-hand corner at $(0, 0)$, lower-right-hand
corner at $(W, 0)$, and so on, the tiling of Figure 2.4(b) can be represented by

$$((0, 0), 11), \quad ((10, 11), 1), \quad ((11, 0), 12), \quad ((23, 0), 6),$$
$$((23, 6), 6), \quad ((0, 11), 10), \quad ((0, 21), 10), \quad ((10, 12), 19).$$

This representation makes it easy to define a linear order on the set U. Let
us say that

$$(X, Y) \text{ precedes } (X', Y') \text{ if and only if } \begin{cases} X + Y < X' + Y', \\ \qquad\qquad \text{or} \\ X + Y = X' + Y' \quad \text{and} \quad Y < Y', \end{cases}$$

and let us also impose some arbitrary order on the set of squares. Then we
shall say that

$((X, Y), Z)$ precedes $((X', Y'), Z')$

$$\text{if and only if } \begin{cases} (X, Y) \text{ precedes } (X', Y') \text{ in the set of points,} \\ \qquad\qquad \text{or} \\ (X, Y) = (X', Y') \text{ and } Z \text{ precedes } Z' \text{ in the set of squares.} \end{cases}$$

The application of the backtrack algorithm of Section 2.1 corresponds to a systematic search for an arrangement of the squares, which tries to fill the available space from the bottom to the top and from the left to the right. The details of the requirements a set of pairs must satisfy are straightforward and are left for the reader to discover.

2.4. GRAPH ALGORITHMS

Graphs are used to represent a wide variety of objects: road systems, electrical networks, the organizational structure of a society, the timing constraints which exist among various activities that are part of a large process—these are only some of the applications of graphs which have become increasingly popular in recent years. It is fair to say that, from the point of view of applications, graph theory is the most important part of combinatorial mathematics. Hence anybody interested in computer applications should know some of the important concepts and some typical algorithms on graphs.

A *graph* consists of a finite set of points, called the *nodes* or *vertices* of the graph, and a finite set of *edges*, or *branches*, each one of which connects a pair of nodes. The graph shown in Figure 2.5 has four nodes labeled 1, 2, 3, and 4, and five edges labeled *a*, *b*, *c*, *d*, and *e*.

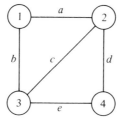

Fig. 2.5 An example of a graph with four nodes and five edges.

There are many variations on the definition of a graph. For instance, one may or may not allow edges whose end points are the same node (self-loops), or more than one edge between two given nodes (multiple or parallel edges). Sometimes the edges are *directed*, so that an edge goes *from* node *i* *to* node *j*, in which case we speak of a directed graph, or digraph, for short.

There are many ways to represent graphs. Pictures are most convenient for humans (at least if the graph is not too large), but for machine processing other representations are more useful. Let us describe two representations based on matrices, and an arbitrary numbering of nodes and edges. An *incidence matrix* M of a graph G with n nodes and m edges has n rows (one for each node) and m columns (one for each edge). Matrix element M_{ij} is 1 if node *i* is incident with edge *j* (i.e., edge *j* has node *i* as one of its end points), and 0 otherwise. If G has no self-loops, every column of M has exactly two 1's. A self-loop shows itself as a column with a single 1, and parallel or multiple edges give rise to identical columns.

Frequently, one works only with graphs that have no parallel edges (though later we shall discuss an algorithm for constructing all spanning trees on a graph, where parallel edges arise naturally during the computation even if there were none in the original graph). Such graphs are conveniently represented by an *adjacency matrix*. For a graph with n nodes, this is an n by n matrix, A, of 0's and 1's; the element A_{ij} is 1 if there is an edge from node numbered i to node numbered j, and it is 0 otherwise. Notice that because of the arbitrary ordering of nodes, a graph may have many different adjacency matrices. For the graph of Figure 2.5, the two orders 1, 2, 3, 4 and 1, 3, 2, 4 lead to

	1	2	3	4
1	0	1	0	1
2	1	0	1	1
3	0	1	0	1
4	1	1	1	0

or

	1	3	2	4
1	0	0	1	1
3	0	0	1	1
2	1	1	0	1
4	1	1	1	0

as adjacency matrices. In general, if A is an adjacency matrix of a graph G, any permutation applied simultaneously to both the rows and the columns of A is an adjacency matrix of the same graph.

This section will study several algorithms that operate on graphs. These have been chosen to illustrate different techniques and to introduce some of the major types of graph algorithms. The shortest-path algorithm is an example of a *propagation* algorithm. One begins by looking at one node and then successively considers all neighbors of this node, then all neighbors of neighbors, and so on, and thus scans a graph in a way similar to the way in which a wave would propagate through the graph. Warshall's algorithm for computing the connected components of a graph and its generalization for computing the distances between all pairs of nodes show two useful matrix operations. The algorithm for constructing a minimal-cost spanning tree may be said to be of the *partition* type, where one must partition the set of nodes or the set of edges into classes so as to satisfy certain constraints. Typically, such algorithms proceed by making a trial assignment to various classes and then successively exchanging elements that violate some constraint until all conditions are met. Finally, the algorithm for generating all spanning trees on a graph is an instance of a *decomposition* algorithm. Decomposition algorithms attempt to answer a question about a graph by decomposing it into smaller graphs and recursively asking the same question for each of the smaller graphs. It is assumed that this question can be answered directly for graphs of sufficiently small size.

Many, but of course not all, graph algorithms fall naturally into one of the classes described here.

2.4.1. Shortest Paths Between Two Nodes

A path is a sequence $N_0, N_1, \ldots, N_l$ of distinct nodes in a graph G such that there is an edge e_i with end points N_{i-1} and N_i for $i = 1, \ldots, l$. We call this a path from N_0 to N_l. The length of a path is just the number l of edges involved; the distance between two nodes in a graph is the length of the shortest path(s) between them.

We shall now describe an algorithm that, given an arbitrary graph and two of its nodes X and Y, answers the following questions:

1. What is the distance from X to Y?
2. How many distinct shortest paths are there from X to Y?
3. What are all the shortest paths from X to Y?

There is a point that deserves mentioning before we describe the shortest-path algorithm. It is necessary in many problems about graphs to specify in what form the answer is to be given. Clearly, for questions 1 and 2 we want a number as the answer, but it is not clear in what form we want all the shortest paths from X to Y. If there are many shortest paths, the most compact representation of all of them just might be the graph itself, and not, say, a node-by-node listing of every path. We shall show how to provide the answer in the form of the original graph with all edges removed that do not lie on any shortest paths, and with an arrow attached to every remaining edge, so if one follows arrows one is guaranteed to travel on a shortest path.

The shortest-path algorithm to be described proceeds by systematically tracing all paths of increasing length from one of the nodes, say X. By marking each node with the distance it has from X, one avoids tracing the paths that double back upon themselves or contain a circuit. The process stops when the goal node Y has been reached, or when no path can be extended farther. In the latter case, there is no path from X to Y; in the former case, a shortest path can be traced by going backward from Y in any way which satisfies the condition that the distance from X decreases by 1 at each step. An explicit description is as follows:

Step 1 (Initialize)

Set $k \leftarrow 0$ and set $WF \leftarrow \{X\}$. WF is the "wavefront" whose value is a set of nodes.

Step 2 (Mark nodes)

Mark all the nodes in WF with k. If WF is empty, then stop; there is no path from X to Y. If $Y \in WF$, go to step 3 (the distance from X to Y is k). Otherwise, increment k by 1, set $WF \leftarrow \{$all unmarked neighbors of nodes in $WF\}$, and repeat step 2.

Step 3 (Initialize for retrace)
    Set $RWF \leftarrow \{Y\}$.

Step 4 (Retrace)
    Decrement k by 1 and set $M \leftarrow$ {nodes marked with k}, and mark
    with an arrow all those edges leading from a node in RWF to a
    node in M. If $k = 0$, stop; otherwise, set $RWF \leftarrow$ {nodes in M that
    are incident to marked edges} and repeat step 4.

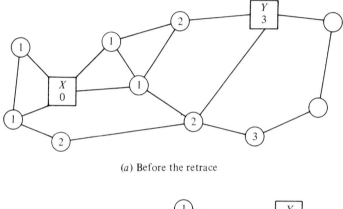

(*a*) Before the retrace

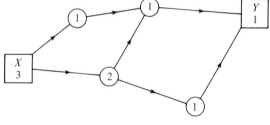

(*b*) After the retrace

Fig. 2.6   A graph as marked by shortest-path algorithm.

On a graph with nodes and edges thus marked, one can travel a path of
shortest length from X to Y by moving along edges marked with arrows,
always moving in the same direction with respect to the arrows. Figure 2.6
shows a graph with nodes and edges marked by this algorithm.

Counting the number of distinct shortest paths can be done by the
following systematic process, which can be conveniently incorporated into
the "Retrace" part of the previous algorithm:

1. Attach as a second label to the node Y the number 1; this has the
intuitive meaning that there is one shortest path from Y to Y.

2. To every other node z that lies on a shortest path, attach as a second
label the sum of the second labels of all nodes which can be reached from z

in one step by following an arrow. Figure 2.6(b) shows the subgraph of all shortest paths of the graph in Figure 2.6(a), the nodes now being marked with these second labels. The label 3 computed for X indicates that there are three distinct paths from X to Y.

2.4.2. Connectedness and the Distances Between All Pairs of Nodes

A graph is said to be *connected* if every pair of nodes is joined by some path. A maximal connected subgraph of a graph is said to be a connected component. Figure 2.7 shows a graph with seven nodes that has two connected

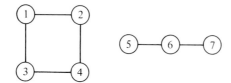

Fig. 2.7 A graph with two connected components.

components. Its adjacency matrix A is

	1	2	3	4	5	6	7
1	0	1	1	0	0	0	0
2	1	0	0	1	0	0	0
3	1	0	0	1	0	0	0
$A = $ 4	0	1	1	0	0	0	0
5	0	0	0	0	0	1	0
6	0	0	0	0	1	0	1
7	0	0	0	0	0	1	0.

The connectivity matrix C of a graph with n nodes is an n by n matrix in which C_{ij} is 1 if there is a path that connects nodes i and j, and 0 otherwise. For the graph in Figure 2.7 this is

	1	2	3	4	5	6	7
1	1	1	1	1	0	0	0
2	1	1	1	1	0	0	0
3	1	1	1	1	0	0	0
$C = $ 4	1	1	1	1	0	0	0
5	0	0	0	0	1	1	1
6	0	0	0	0	1	1	1
7	0	0	0	0	1	1	1.

Clearly, if the nodes have been numbered appropriately, the connected components show up as square submatrices of 1's.

The problem we wish to consider is how the connectivity matrix C can be computed efficiently from the adjacency matrix A. To express algorithms conveniently, we introduce two operations $\wedge$ (AND) and $\vee$ (OR) on variables that can assume only the values 0 and 1. These operations are defined by the following tables:

$$x \wedge y: \quad \begin{array}{c|cc} {}_x\diagdown{}^y & 0 & 1 \\ \hline 0 & 0 & 0 \\ 1 & 0 & 1 \end{array} \qquad x \vee y: \quad \begin{array}{c|cc} {}_x\diagdown{}^y & 0 & 1 \\ \hline 0 & 0 & 1 \\ 1 & 1 & 1. \end{array}$$

An elegant algorithm due to Warshall finds C by computing a sequence of n by n matrices $B^0, B^1, \ldots, B^n$ according to the following rules:

Warshall's Algorithm

1. $B^0 \leftarrow A$.
2. For $l = 1, 2, \ldots, n$, compute B^l as follows:

$$B_{ij}^l \longleftarrow B_{ij}^{l-1} \vee [B_{il}^{l-1} \wedge B_{lj}^{l-1}].$$

3. $C \leftarrow B^n$.

It is easy to see that this algorithm is correct if one interprets the quantities B_{ij}^l appropriately. We want B_{ij}^l to be 1 if and only if the nodes i and j are connected by a path that has as intermediate nodes only nodes chosen from the set $\{1, 2, \ldots, l\}$. Notice that the initialization step $B^0 \leftarrow A$ (paths with no intermediate nodes) and the final step $C \leftarrow B^n$ (paths with any intermediate nodes whatever) are consistent with this interpretation. And the formula in the iterative step merely expresses the fact that for i and j to be connected by a path which has intermediate nodes only in the set $\{1, 2, \ldots, l\}$ it is necessary and sufficient that one or both of the following conditions hold:

1. There is a path from i to j which has intermediate nodes only in the set $\{1, 2, \ldots, l - 1\}$.
2. There is a path from i to l and a path from l to j, both of which have intermediate nodes only in the set $\{1, 2, \ldots, l - 1\}$.

Figure 2.8 illustrates the second condition (you should verify that the argument does not change if nodes i and/or j are in the set $\{1, 2, \ldots, l\}$).

Warshall's algorithm can be generalized to compute the *distance* between all pairs of nodes in a graph. Here we shall also generalize the notion of distance between two nodes in a graph, from simply counting the number of edges in a path to summing the *lengths* of edges. We assume each edge in a

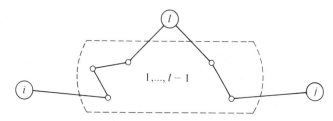

Fig. 2.8 One of the conditions in Warshall's algorithm.

graph is assigned a nonnegative real number, called the length of this edge. The length of a path is now the sum of the lengths of all edges in the path. This notion specializes to the previous one if all edges are assigned unit length. It is interesting to note that we do not assume the edge lengths satisfy any "triangle inequality"; the shortest path between two adjacent nodes may very well contain many short edges, instead of one long edge.

Let the matrix A now indicate the length of an edge instead of being the adjacency matrix, which merely indicates existence or nonexistence of an edge between two nodes. We shall use the convention that

$$A_{ij} = \begin{cases} \text{length of edge between } i \text{ and } j, \text{ if it exists,} \\ \infty \text{ if there is no edge between nodes } i \text{ and } j. \end{cases}$$

We shall carry out the arithmetic operations of addition and minimization on the symbol ∞ in the obvious way: if x is any number, then $x + \infty = \infty$ and $\min(x, \infty) = x$.

We want to compute, from this matrix A, which is a generalization of the adjacency matrix, a distance matrix D, which is a generalization of the connectivity matrix C considered earlier. D is defined by

$$D_{ij} = \begin{cases} \text{length of shortest path between } i \text{ and } j, \text{ if it exists,} \\ \infty \text{ if there is no path from } i \text{ to } j. \end{cases}$$

An analogous interpretation of the quantities B_{ij}^l should convince the reader that the following algorithm correctly computes D from A.

1. $B^0 \longleftarrow A$.
2. For $l = 1, 2, \ldots, n$, compute B^l as follows:

$$B_{ij}^l \longleftarrow \min[B_{ij}^{l-1}, B_{il}^{l-1} + B_{lj}^{l-1}].$$

3. $D \longleftarrow B^n$.

2.4.3. Minimal-Cost Spanning Trees

A *circuit*, or closed path, is a path as defined previously, except that the first and last nodes are equal; a *tree* is a graph with no circuits. A subgraph G' of G that is a tree and that is maximal (no edge of G can be added to it without forming a circuit) is called a *maximal tree* or a *spanning tree* in G. The following properties hold for a spanning tree T of a graph G:

1. Every node of G is incident to at least one edge of T.

2. The number of edges in T is equal to the number of nodes in G minus the number of connected components of G (and hence equal to $n - 1$ for a connected graph of n nodes).

The problem of finding spanning trees can be interpreted as looking for economical ways of connecting a given set of points to each other without duplicating connections, and so it has applications in various fields, such as the study of electrical networks or the analysis of the frequency with which branches in flowcharts are taken. Consider a graph G each of whose edges e has been assigned a number $c(e)$, called the weight or cost of e. The *cost of a spanning tree* is the sum of the costs of all the edges in the tree. We want to find a spanning tree whose cost is minimal.

Our problem then is to partition the set of edges of G into two classes, subject to the restrictions that the branches in one class form a spanning tree and that their total cost is not greater than the total cost of any other spanning tree. Assume that we have a spanning tree to begin with; our algorithm will go through a sequence of *exchange steps*, in which an edge not currently in the tree is added to the tree and an appropriate edge in the old tree is deleted. The algorithm to be discussed is rather unusual among combinatorial algorithms in that it leads directly from any initial spanning tree we may have chosen to one of minimal cost. When no exchange step can be performed, we are guaranteed to have found a minimal-cost spanning tree. In the space of all spanning trees, there are no local minima from which one has to "back up" in order to reach an absolute minimum.

It is useful, whenever one deals with spanning trees, to introduce a function $CIR(e, T)$, which has as arguments a set of edges that form a spanning tree T, and a single edge e not in T. The value of $CIR(e, T)$ for these arguments is the set of all edges in T which lie on the unique path in T that connects the two end points of e; see Figure 2.9 for examples. In terms of this function the following algorithm can be expressed concisely.

Exchange Algorithm

1. Let T be any spanning tree in G.
2. Examine in turn all edges e not in T, and for each one do the following:

    (a) Find an edge e' of maximal cost in $CIR(e, T)$.

    (b) If $c(e) \geq c(e')$, do nothing, but if $c(e) < c(e')$, modify T by deleting e' and replacing it by e (we call this a cost-reducing exchange and express it as $T \leftarrow T - e' + e$).

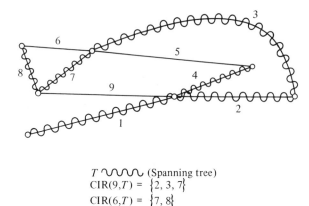

T ⟆⟆⟆⟆ (Spanning tree)
CIR$(9,T)$ = $\{2, 3, 7\}$
CIR$(6,T)$ = $\{7, 8\}$
CIR$(5,T)$ = $\{2, 3, 4\}$

Fig. 2.9   Examples of the function CIR(e, T).

3. Stop when there are no more edges e outside T that can be exchanged for an edge e' in CIR(e, T) with an improvement in cost.

It is clear that the cost of the final tree T found by this algorithm is not greater than the cost of any of the intermediate trees constructed by the algorithm. T is in fact of minimal cost among all spanning trees. In particular, all the trees constructed by this algorithm have the same cost, independent of the initial spanning tree used. The proof of this fact is best presented by way of contradiction.

Assume that the tree T constructed by the exchange algorithm is not a minimal-cost spanning tree; in other words, it is a *local* minimum in the sense that no cost-reducing exchanges are possible. Then there exists a tree, U, in the set of global minimal-cost spanning trees which is closest to T in the sense that it has as many edges in common with T as possible. Consider the set of edges in T but not in U. By assumption, this set is nonempty and hence has a *least-cost* edge, say e. (See Figure 2.10.) Since U is a spanning tree, adding the edge e to U forms a circuit, CIR(e, U), which must contain an edge of U that is not in T (i.e., it is in $U - T$). We claim that for every edge in u in CIR(e, U), $c(u) < c(e)$. Otherwise, by performing the exchange $U' \leftarrow U - u + e$, one could either improve the cost of U, contradicting the assumption that U is minimal, or obtain another minimal-cost spanning tree U' closer to T than U. Now let u be an edge of $U - T$ in CIR(e, U), and consider the circuit CIR(u, T). As above, this circuit must contain an edge e' not in U; but then by the fact that T allows no cost-reducing exchanges $T \leftarrow T - e' + u$, we have $c(e') \le c(u)$, and hence $c(e') < c(e)$, contradicting the least-cost property of e in $T - U$. Thus we must conclude that $T - U$ is in fact empty, and so T is a minimal-cost spanning tree.

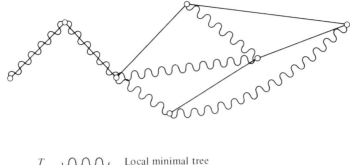

T ∿∿∿   Local minimal tree

U ────────   Global minimal tree

Fig. 2.10   Local versus global minimal spanning trees.

One possible shortcoming of the preceding algorithm for generating a minimal-cost spanning tree is the assumption at the beginning that we already have at least one spanning tree. In some applications, such as analysis of electrical networks, it is convenient to choose the representation of a graph so as to explicitly contain a spanning tree (see Exercise 13). On the other hand, if one has to construct a spanning tree first, one can construct one of minimal cost almost as easily. An obvious method for finding an arbitrary spanning tree is to construct a set of edges one at a time in such a way that when a new edge is added no circuit is formed with the set of edges already constructed. What modifications of this procedure could be made so that one is guaranteed to have a minimal-cost spanning tree? The following algorithm illustrates what changes might be made.

Algorithm 1

Repeat the following step as long as possible: select from the edges not already chosen an edge of *least cost* that will not form a circuit with the set already chosen.

The final set of edges chosen in this way forms a minimal-cost spanning tree.

Algorithm 2 is, in a sense, the dual to Algorithm 1:

Algorithm 2

Beginning with an entire connected graph G, repeat the following step as long as possible: remove an edge of *highest cost* that will leave the remaining edges connected.

The set of edges remaining after no more can be removed forms a minimal-cost spanning tree.

The proofs that these two algorithms actually yield minimal-cost spanning trees are not difficult (using the fact, proved previously, that a tree which allows no cost-reducing exchanges is of minimal cost), and are left as exercises for the reader. Which one of the three algorithms is the most efficient depends on the representation in which the graph is given.

Let us emphasize again that for graph problems it is rather unusual that algorithms as simple as these exist for finding a subgraph which is minimal in a certain sense. More typical of the difficulties that may be encountered is the "traveling-salesman problem": a salesman has to visit each of n cities exactly once and return to his starting point; given the distances between each pair of cities, find the shortest route. This problem differs from ours in that, instead of a minimal-cost spanning tree, what is needed is a minimal-cost *Hamiltonian circuit*, that is, a circuit that contains every node of a graph exactly once. It turns out, however, that this problem is computationally much more difficult than the one about spanning trees. In contrast to the exchange algorithm described previously, if we start with an arbitrary Hamiltonian circuit and make some local modifications that yield a cheaper circuit, then when no cost-reducing modifications can be made, we have found only a *local minimum*, not necessarily a global one.

2.4.4. Finding All Spanning Trees of a Graph

The problem of finding all spanning trees of a graph arises in topological methods for analyzing linear electrical networks. Because of its importance, at least a dozen different algorithms have been proposed for solving this problem.

The enumeration of all spanning trees of a graph looks like a straight-forward application of the backtrack procedure. By imposing an arbitrary order on the set of edges or the set of nodes of the underlying graph, backtrack will automatically assure that no tree is overlooked and no tree is generated more than once. Indeed, most algorithms for generating all spanning trees are based on backtrack. This explains the multiplicity of algorithms, since different formulations of the problem in terms of backtrack may lead to significantly different computation times.

The number of spanning trees on even small graphs is large. A theorem of Cayley states that on a *complete graph* of n nodes (a graph is complete if every pair of nodes is joined by an edge) there are n^{n-2} distinct spanning trees. Thus a complete graph of 10 nodes has 10^8 distinct spanning trees, and most graphs of 10 nodes still have sufficiently many spanning trees to make the problem of finding them all nontrivial, even on a fast computer.

We shall present only the central idea of some of the faster algorithms for generating all spanning trees, because it introduces an elegant and powerful method: to solve a problem about a large graph by decomposing the graph into several smaller ones, and recursively solving the same problem for each of the small graphs. To ensure that the procedure terminates, sufficiently simple graphs are not decomposed further, but instead the problem is solved directly on them.

To apply this principle to the generation of all spanning trees, consider a graph G and one of its edges, e. The set T of all spanning trees on G can now be partitioned into two classes, T', consisting of those trees which do not contain e, and T'', consisting of those trees which contain e. It is obvious that T' is just the set of all spanning trees on the graph G', obtained by removing edge e from G. The set T'' can be put into one-to-one correspondence with the set of all spanning trees on a graph G'', obtained from G by merging the two end points x, y of e in G into a single new node z in G''. All edges that were incident to x or y in G are now incident to z in G''. Any self-loops that may have been created through this merging are dropped in G''; on the other hand, multiple edges must be kept in G''.

Notice that the two graphs G' and G'' derived from G are smaller than G, since they contain fewer nodes or edges. Hence this algorithm must eventually terminate. The formulation of an efficient criterion of when a graph is sufficiently simple not to be further decomposed is left as an exercise to the reader.

Figure 2.11 shows how this algorithm finds all three spanning trees of the complete graph of three nodes. The procedure terminates when a graph splits into two components or consists of a single node. Beginning with a graph G, the left arrow points to the graph G' derived by edge deletion, and the right arrow to G'' derived by node merging. Notice that the information about what was merged must be retained in order to be able to reconstruct the spanning tree of the original graph.

2.5. SORTING

One of the most frequently performed tasks on computers is a combinatorial problem that the average person, unacquainted with computers, is likely to regard as trivial: sorting, that is, arranging items according to a predefined order. However, sorting is far from trivial, both from the practical and the theoretical aspects. Considerable effort has been expended on problems related to sorting; in this section we describe some of the central ideas and results so that the reader can develop an appreciation of the importance and the difficulty of the subject.

In conventional data-processing terminology, one sorts a *file* of *records* according to their *key*. An example will explain the meaning of these terms. A company maintains a *file* of its employees. All the data for each individual employee forms a *record*, which contains several fields (pieces of informa-

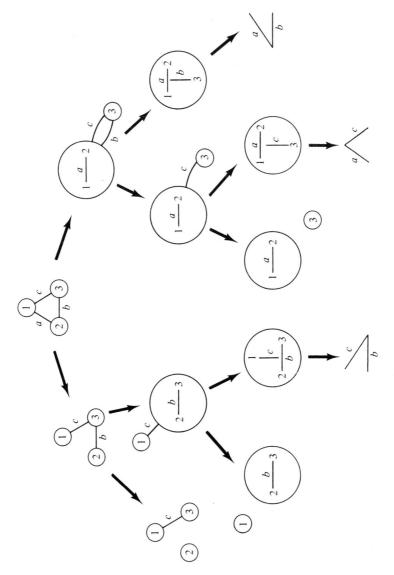

Fig. 2.11 Generating all spanning trees of the complete graph of three nodes.

tion): say, his or her name, social security number, job category, salary, and so forth. For a sorting operation, one of these fields is considered to be the *key*. If the name field is the key, we sort the file in alphabetical order of the names. If the social security number is the key, we sort in increasing numerical order of the social security numbers. For our purpose the only important aspect is that the set of keys is (linearly, or totally) ordered.

To simplify matters, we shall assume that the keys are always integers and that they are initially presented in an unknown order $x_1, x_2, \ldots, x_n$. By sorting we mean the rearrangement of this sequence x_i into ascending order based on certain observations of and operations on these elements. It is convenient to choose as the unit of observation a comparison of two elements, x_i and x_j. Each such comparison yields as its result a binary outcome: either $x_i < x_j$ or $x_i > x_j$ (for the sake of convenience, we shall exclude the third possibility, $x_i = x_j$; it can easily be taken into account—see Exercise 16).

As a unit of operation we shall consider either a transposition of two elements x_i and x_j, in which each one takes the place of the other, or the moving of a single element from one position and inserting it as a left or right neighbor of another element. Which of these two units of work we consider depends on the data structure used. If the elements are located in a vector of n fixed positions, as shown in Figure 2.12(a), transpositions are most meaningful as the unit of work. On the other hand, if the elements are linked together by pointers in a linked linear list, as shown in Figure 2.12(b), insertions are most meaningful.

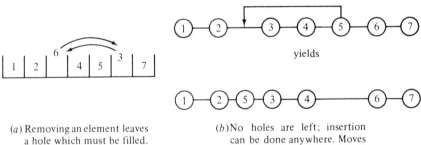

(a) Removing an element leaves a hole which must be filled. Transpositions are natural in this case.

(b) No holes are left; insertion can be done anywhere. Moves are natural in this case.

Fig. 2.12   The applicability of transpositions and moves as units of work.

Generally speaking, the sorting algorithms admitted by our model fall into one of three important categories. First, there are *transposition sorts*. In this type of sort, pairs of keys are examined, and if the keys are out of order, they are interchanged or transposed; this process is continued until all the keys are in order. Next, there are *selection sorts* in which the keys are selected one at a time in increasing order and stored in order as they are selected; thus the smallest is found, then the next smallest, and so on. The third type of sort is known as an *insertion sort*. Here the keys are examined one by one, each key being inserted into its proper location relative to the keys previously examined. These three classes of sorting algorithms are illustrated in Figure 2.13. Notice that in the last two types of sorting algorithm the set of keys to be sorted is partitioned into two classes, those already sorted and those not yet sorted; initially, the sorted class is empty and the unsorted class contains all the keys, whereas at the termination of the algorithm the unsorted class is empty and the sorted class contains all the keys. The algorithms iterate a step in which one element is transferred from the unsorted class to the sorted class.

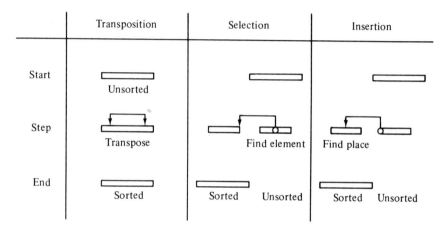

Fig. 2.13   Three classes of sorting algorithms.

Our model of sorting is simplified in several respects. For one thing, some methods of sorting are based on a knowledge of the statistical distribution of the keys to be sorted, or on the particular representation of these keys†; our model will allow no such techniques. The simplification which is most crucial from a practical point of view is that we demand instant access to all keys in the file to be sorted. This is known as *internal sorting*, when all records (or at least all keys, each containing a pointer to its record) fit into the main memory. Usually, however, files are so large that only a small part fits into main memory; the bulk is kept on backup storage devices with limited access, from which keys can be transferred to main memory, but only in a sequential manner. This is known as *external sorting*; it is even more difficult to

---

†For example, knowing that the keys are the integers from 1 to n clearly adds a lot of information as compared to only knowing that the keys are arbitrary integers. To see this, consider what can be said about the position of any key in the sorted file.

analyze than internal sorting, because one must make assumptions about the properties of backup storage devices. We shall consider only internal sorting.

The available knowledge and theory about internal sorting fall into two categories: discussions of specific sorting algorithms on the one hand, and the principles and properties common to all sorting algorithms on the other. The remaining sections give insight into these areas.

2.5.1. Transposition Sorting

As mentioned in the previous section, a transposition sort is one that selectively interchanges pairs of elements which are out of order. One of the simplest such sorts is known as the *bubble sort*, so named because it operates in analogy with the way bubbles rise to the top in a carbonated liquid: the "lighter" elements "bubble up" to the top part of the file. First, the second key is compared with the first and interchanged if necessary. Then the third key is compared with the second, and if an interchange is made it is compared with the first key. This continues as each key in turn is compared with those

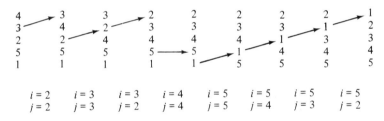

Fig. 2.14 An example of the bubble sort in action.

above it, as illustrated in Figure 2.14. The details of the algorithm are

Repeat for i from 2 through n, incremented by 1:
    Repeat for j from i through 2, decremented by 1, or until $x_j \geqq x_{j-1}$:
        Interchange x_{j-1} and x_j.

It is clear that in the *worst case*, when the keys are in exactly the reverse order, the bubble sort requires about $\frac{1}{2}n^2$ comparisons and interchanges. What is more important, however, is that it can be shown that in the *average case* (assuming each of the $n!$ permutations is equally probable) about $\frac{1}{4}n^2$ comparisons and interchanges are required. As we shall see in subsequent sections, there are sorting algorithms which require only about $n \log n$ operations, so that the bubble sort is relatively inefficient. Its only real advantage is common to all transposition sorts: it uses almost no memory outside the area in which the keys are stored—it is an *in-place* sorting algorithm. A minor advantage which it enjoys over other algorithms is that it is conceptually simple and easy to program.

The reason the bubble sort is so slow is that interchanges and comparisons are made only between adjacent keys. To get a transposition sort that is significantly faster than the bubble sort, comparisons and interchanges must be made between keys which are widely separated. One application of this idea results in a sorting algorithm known as Quicksort. Quicksort operates by using the first key in the unsorted file as a guess at the median of the set of keys. Then, based on that guess, the keys are split into two groups, those greater than the guess and those less than it; the two groups are then sorted separately, using the same Quicksort idea. This can be done in place by scanning down from the top looking for keys greater than the guess and scanning up from the bottom looking for keys less than the guess. When a greater key above and a lesser key below are found, they are interchanged; finally, when the scans meet, the guess is interchanged so that it is between the two groups. This process is illustrated in Figure 2.15.

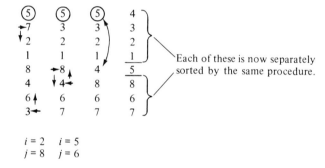

$i = 2 \quad i = 5$
$j = 8 \quad j = 6$

Fig. 2.15 The first stage of Quicksort. The variables i and j are explained in the Quicksort algorithm.

The Quicksort algorithm is conceptually easy to understand, but writing a program for it requires some thought if the algorithm is to be iterative rather than recursive. The recursive form of the algorithm is given next; we leave the iterative version to Exercise 17.

Step 1

If $n \leq 1$, the file is sorted, so stop. Otherwise, initialize by setting $i \leftarrow 1$ and $j \leftarrow n$; i starts at the top and scans down while j starts at the bottom and scans up.

Step 2

If $x_j \leq x_1$ go on to step 3, otherwise set $j \leftarrow j - 1$ and repeat step 2.

Step 3

> If $i = j$ go to step 5, otherwise set $i \leftarrow i + 1$. If $x_i < x_1$ repeat step 3, otherwise go on to step 4.

Step 4

> Interchange x_i and x_j and go back to step 2.

Step 5

> The two scans have met; interchange x_1 and x_i.

Step 6

> Using the same algorithm, separately sort the files $(x_1, \ldots, x_{i-1})$ and $(x_{i+1}, \ldots, x_n)$. Notice that if $i = 1$ the former file is empty, and if $i = n$ the latter file is empty.

Is Quicksort an improvement over the bubble sort? At first sight it might seem that the minor change in the order in which the keys are interchanged only bought us a considerably more difficult program to write! In the worst case, which occurs either when the elements are in reverse order or when they are in the correct order (!), Quicksort requires proportional to n^2 operations. However, Quicksort requires, on the average, only about $2n \log n$ comparisons and fewer interchanges, so it does represent an improvement over the bubble sort, particularly since the average work required is certainly a more reasonable measure than that which is required in the pathological cases. Quicksort is not good in the worst case, because the guesses it makes for the median are so extremely poor.

2.5.2. Selection Sorting

Selection sorting is based on the repeated selection of the smallest key of the keys to be sorted. First, we find the minimum key and transfer it to the output; then we find the minimum key of those remaining, and transfer it to the output. The process continues until all the keys are in the output. This idea can be combined with that of the transposition sort to yield the following simple algorithm to sort $\{x_1, \ldots, x_n\}$:

Step 1

> Initialize by setting $i \leftarrow 0$.

Step 2

> Set $i \leftarrow i + 1$. If $i = n$, we are finished.

Step 3

> Find the smallest of $\{x_i, \ldots, x_n\}$, interchange it with x_i, and return to step 2.

How efficient is this method? It depends on how fast we can find the smallest of i keys. The obvious left-to-right scan requires $i - 1$ comparisons. Since this must be done for $i = n, n - 1, \ldots, 1$, step 3 alone will always require $\sum_{i=1}^{n} (i - 1) \approx \frac{1}{2}n^2$ comparisons and $n - 1$ interchanges. Certainly then, as constituted, this algorithm is no improvement on the sorting algorithms given in the previous sections; but, can we modify the algorithm to make it faster? The bottleneck is obviously the operation of finding the smallest element each time at step 3. Is there a method for finding the minimum of a set of i keys in fewer than $i - 1$ comparisons? No, in fact it can easily be shown that

Theorem

Given the set of keys $\{x_1, \ldots, x_i\}$, $i - 1$ comparisons are necessary (and, of course, sufficient) to find the smallest key.

The proof is left as Exercise 18.

This theorem seems to say that selection sorting is doomed to always require about $\frac{1}{2}n^2$ comparisons. Fortunately, this is not the case, for although it is true that finding the *first* minimum will require $n - 1$ comparisons, much information from these comparisons can be used in finding the second minimum. Consider the usual knock-out tournament, as illustrated in Figure 2.16.

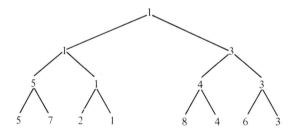

Fig. 2.16   A knock-out tournament of find the minimum.

The $n - 1$ comparisons used to find the "winner" (in our case, the smallest key) are arranged in the form of a tree. At each level of the tree the numbers are paired off, the smaller of a pair being the winner. The overall winner is at the root of the tree and must be the smallest key. We put that element into the output and go on to find the next smallest key; to do that, we delete the winner from the tournament at the bottom of the tree, replacing it with ∞, and then remake the comparisons from that ∞ to the root of the tree. The result is illustrated in Figure 2.17. Now we know that 2 is the second smallest key, and we can put it into the output and replace it in the tree with ∞, remarking the comparisons in which it is involved. This process continues until the value ∞ reaches the root of the tree.

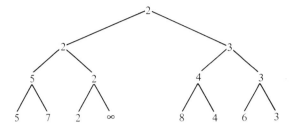

Fig. 2.17 Finding the second smallest element in a knock-out tournament.

The number of comparisons needed for the entire sort is easily calculated. The first stage of the knock-out tournament requires $n - 1$ comparisons in addition to some small amount of "setup" time. Since there are n leaves in the tree, there will be about $\frac{1}{2}n$ winners at the first level, about $\frac{1}{4}n$ winners at the second level, and so on; at the ith level there will be about $n/2^i$ winners. Clearly, the root of the tree is encountered when $0 < n/2^i \leq 1$, or when $i \approx \log_2 n$. Thus, after finding the minimum, only about $\log_2 n$ of the comparisons must be recomputed to find the second, third, and remaining minima. Since there are $n - 1$ minima to be found after the first, about $(n - 1)(1 + \log_2 n)$ comparisons are used altogether. Thus the *tournament-selection sort* just described requires at most an amount of work proportional to $n \log n$.

2.5.3. Insertion Sorting

The basic technique of insertion sorting is one that is likely to be used by humans when they do manual sorting: take the keys one at a time and insert them into their correct place relative to the already sorted part of the file. This complements the selection technique in the previous section; there we selected keys in order; here we order them as they are selected. The simplest insertion sorting algorithm is

For j from 2 to n incremented by 1, perform the following three steps:

Step 1
> Set $i \leftarrow j - 1$.

Step 2
> If $x_j \geq x_i$, go on to step 3; otherwise, decrement i by 1; then if $i \geq 0$, repeat step 2.

Step 3
> At this point we have found the place where x_j should be inserted (just before x_{i+1}). Set $t \leftarrow x_j$ and then for $k = j, j - 1, \ldots, i + 1$, set $x_k \leftarrow x_{k-1}$. Finally, set $x_i \leftarrow t$.

In step 2 of this algorithm we expect, on the average, to compare x_j with about half the $j - 1$ previously inserted keys; thus the average number of comparisons made by the algorithm is $\sum_{j=2}^{n} \frac{1}{2}(j - 1) \approx \frac{1}{4}n^2$; of course, in the worst case about $\sum_{j=2}^{n} (j - 1) \approx \frac{1}{2}n^2$ comparisons are used. We can substantially improve this part of the algorithm by replacing steps 1 and 2 with

Step 1
   Set $l \leftarrow 1$ and $u \leftarrow j$.

Step 2A
   If $u \leq l$, set $i \leftarrow u$ and go to step 3; otherwise, set $m \leftarrow [(l + u)/2]$.†

Step 2B
   If $x_j < x_m$, set $u \leftarrow m - 1$; otherwise, set $l \leftarrow m + 1$. Go to step 2A.

Although this is considerably more complicated, it saves a great deal of time, for what we have done is replace the *linear search* in the original algorithm with a *binary search*. Whereas the linear search examines n keys one at a time from right to left, the binary search looks in the middle of those keys to decide which half is of interest. It then examines that half by looking at its middle key to see which half of that half is of interest. The process continues in this manner until we are down to only one key; since with each comparison the number of keys to be examined is cut in approximately half, after i comparisons there are about $n/2^i$ keys left. One key is left when $0 < n/2^i \leq 1$, or when $i \approx \log_2 n$, or after at most about $\log_2 n$ comparisons.

The linear search needed about $\frac{1}{2}(j - 1)$ comparisons, on the average, to find the place to insert the jth key, whereas, as we have seen, the binary search needs only about $\log_2 (j - 1)$ comparisons to do the same thing in the *worst case*. Thus the number of comparisons needed on the average is reduced from about $\frac{1}{4}n^2$ to $\sum_{j=2}^{n} \log_2 (j - 1) \approx n \log_2 n$, and in the worst case the number of comparisons is reduced from about $\frac{1}{2}n^2$ to $n \log_2 n$, a significant improvement. Unfortunately, the insertion algorithm has another drawback: in step 3 approximately $\frac{1}{2}(j - 1)$ moving operations are needed to shift the keys over to make room for the key to be inserted. As before, it obviously requires about $\frac{1}{4}n^2$ moves on the average and about $\frac{1}{2}n^2$ in the worst case, and so the amount of work is still proportional to n^2 for the insertion algorithm.

Actually, we can obviate the problem of shifting keys to make room for an insertion by using a linked-list representation in place of the sequential representation; these two representations are illustrated in Figure 2.18, along with the way the insertion would be done. But we still have not really

---

† $[x]$ denotes the least integer less than or equal to x.

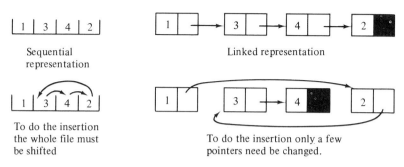

Fig. 2.18   Inserting elements in sequential versus linked files.

improved on the insertion sorting algorithm, since it is impossible to use the binary-search technique on a linked list (see Exercise 20). The next section shows how to improve the insertion sorting algorithm so that it requires work at most proportional to $n \log n$.

2.5.4. Balanced Trees and Sorting

The insertion sorting algorithm can be improved considerably if we abandon the use of a *linear* arrangement of the keys. Rather than store the keys so that each key (except the largest and smallest) has a predecessor and a successor, suppose that we store them so that each key has associated with it two *sets* of keys, one set consisting of *all* predecessors and the other set consisting of *all* successors. The structure best suited for this representation is a *binary-search tree*; Figure 2.19 shows the set $\{1, 2, 3, 4, 5, 6, 7\}$ represented as such a tree. Notice that every node in the tree has the property that all nodes in its left subtree are predecessors and all nodes in the right subtree are successors. Any tree with this property is called a binary-search tree.

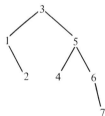

Fig. 2.19   The set $\{1, 2, 3, 4, 5, 6, 7\}$ arranged in a binary-search tree.

Binary-search trees are a popular method for storing and retrieving the keys in a file, because it is relatively easy to examine keys at random or to examine them in sequential order. Suppose that we wish to examine a certain key, say x, in the file. We can do so by the following algorithm, where T is a binary search tree:

Step 1
   If *T* is empty, then *x* is not in the file.

Step 2
   If *x* is equal to the root of *T*, we have found it.

Step 3
   If *x* is less than the root of *T*, replace *T*, by its left subtree, otherwise, replace it by its right subtree. Return to step 1.

On the other hand, if we need to examine the file sequentially, we can do so by
the following recursive "traversal" algorithm:

Step 1
   Traverse the left subtree using this algorithm.

Step 2
   Examine the root.

Step 3
   Traverse the right subtree using this algorithm.

This algorithm has an elegant formulation in an iterative, or nonrecursive
form, using a push-down stack; see Exercise 21.
   Given a binary-search tree, the addition of a new, distinct key is a fairly
simple matter: we use the search algorithm given, and in step 2 when we
find that the key is not in the tree, we add it at that point. Thus to add the
key 0 to the tree in Figure 2.19, we start at the root and compare 0 with 3.
Finding that $0 < 3$, we now go down to the left subtree of 3 and compare 0
with 1. Finding that $0 < 1$, we attempt to go down to the left subtree of 1,
but we find that it does not exist. At this point we add the 0 on as the left
subtree of 1, yielding the tree shown in Figure 2.20.

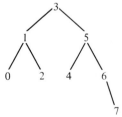

Fig. 2.20 The tree from Fig. 2.19 with
0 added in its proper place.

We now have the basis for a new insertion sorting algorithm: as we
examine the keys we insert them into a binary-search tree, for example,
examining the keys {0, 1, 2, 3, 4, 5, 6, 7} in the order 3, 1, 5, 0, 4, 2, 6, 7 gives
rise to the tree in Figure 2.20. This tree insertion sorting algorithm is very
similar to the Quicksort algorithm described in Section 2.5.1. That algorithm

was, in effect, constructing a binary-search tree for the file, but it was doing it in place. In this algorithm we are constructing the tree in new locations, and the tree is being constructed explicitly instead of implicitly. Like Quick-sort, the tree insertion algorithm can require about $\frac{1}{2}n^2$ comparisons in the worst case, but, on the average, only $n \log_2 n$ comparisons are needed. Certainly, this is an improvement over the plain insertion sorting algorithm discussed earlier, especially since the tree insertion method requires no more than proportional to n operations to do the insertion.

Unfortunately, there is still that annoying worst case in which the work required is proportional to n^2. To eliminate that, we must examine the cause. Suppose that instead of inserting the keys in the order 3, 1, 5, 0, 4, 2, 6, 7 we had inserted them in the order 0, 1, 2, 3, 4, 5, 6, 7. We would have obtained the extremely lopsided tree shown in Figure 2.21, whose construction required about $\frac{1}{2}n^2$ comparisons. A similar problem would have occurred if we had inserted the keys in the order 7, 6, 5, 4, 3, 2, 1, 0. Clearly, the $\frac{1}{2}n^2$ comparisons are needed because the tree has become very "unbalanced"; we can eliminate this problem by "rebalancing" the tree when needed, as the insertions are being made.

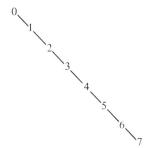

Fig. 2.21 A lopsided binary-search tree for the set $\{0, 1, 2, 3, 4, 5, 6, 7\}$.

One notion of a balanced tree is based on the height of its two subtrees†: the *height* of a tree is the length of the longest path from the root to a leaf. Thus, for example, the tree in Figure 2.21 has height 7, while the tree in Figure 2.20 has height 3; by definition, the height of a tree consisting of a single node is zero. Now, we define a tree to be *height-balanced*, or simply *balanced*, provided that every node in the tree has the property that the heights of the two subtrees of that node differ by at most 1. Figure 2.22 illustrates some balanced and some unbalanced trees.

The problem with the tree insertion sorting algorithm is that if the n keys are inserted in a certain order, the height of the tree becomes propor-tional to n, and the number of comparisons required is proportional to n^2.

---

†There are other reasonable ways to define the notion of a balanced tree; see Exercise 24.

Balanced trees   Unbalanced trees

Fig. 2.22 Examples of balanced and unbalanced trees.

If somehow the height of the tree can be easily kept proportional to $\log n$, then, even in the worst case, $\sum_{j=2}^{n} c \log (j - 1) \approx cn \log n$ comparisons would be needed, thus making the tree insertion an $n \log n$ algorithm, even in the worst case. For this reason we need to show that if a tree is balanced its height must be proportional to $\log n$. Let us turn the problem around and ask: What is the least number of nodes required to form a balanced tree of height k? Let T_k be a balanced tree of height k that contains the fewest possible nodes, say N_k nodes. Now, if T_k is to have the stated property, the left and right subtrees of its root must be (not necessarily respectively) T_{k-1} and T_{k-2}; thus $N_k = N_{k-1} + N_{k-2} + 1$, with $N_0 = 1$, $N_1 = 2$. This recurrence relation is similar to the one satisfied by the Fibonacci numbers (see Section 5.1.3). When solved by the same techniques, it yields

$$N_k = \left(1 + \frac{2}{\sqrt{5}}\right)\left(\frac{1 + \sqrt{5}}{2}\right)^{k-1} + \left(1 - \frac{2}{\sqrt{5}}\right)\left(\frac{1 - \sqrt{5}}{2}\right)^{k-1} - 1.$$

Thus the height of *any* balanced tree with n keys is certainly less than $\frac{3}{2} \log_2 n$, proportional to $\log n$, as desired.

The next thing we must show is that if the insertion of a new key disrupts the balance of a height-balanced tree, the tree can be *rebalanced* easily, that is, so as to keep the total amount of work within the $n \log n$ bound. As it stands, for example, the tree in Figure 2.20 is balanced, but if we tried to insert the key 8, it would become unbalanced at the nodes 6, 5, and 3. This situation always arises when the key to be inserted needs to be put at the lowest point of a subtree that already has one more in height than its "brother" subtree. Aside from reflections, there are two cases that cause trouble, as shown in Figure 2.23; the tree can be rebalanced by applying the corresponding transformation (also shown in that figure) at the lowest point in the tree which is unbalanced (see Exercise 22). Notice that after applying either transformation a tree will still have the property that everything in the left subtree of a key is a predecessor, and everything in the right subtree is a successor.

2.5.5. Sorting Theory

Of all the sorting algorithms we have considered, none has been able to sort in less than $n \log_2 n$ comparisons in the average case, let alone in the worst case. Perhaps we have not been diligent enough, and lurking some-

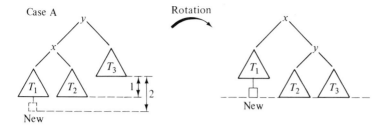

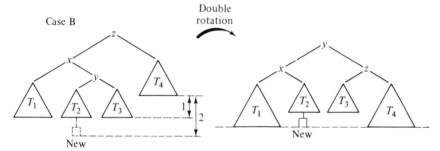

Fig. 2.23 Transformations used to rebalance trees.

where in the unknown is a sorting algorithm that works much faster? No; no matter how hard we work, no sorting algorithm admitted by our model can work any faster than in $n \log_2 n$ comparisons in the worst case or even on the average.

To prove this assertion, notice that any sorting algorithm of the type we consider can be put into the form of a binary tree, each of whose internal nodes corresponds to a comparison between keys in the file, and each of whose edges corresponds to what happens depending on the result of that comparison. Thus the tree is essentially a loop-free flowchart of the sorting algorithm. Figure 2.24 shows how an algorithm that sorts the set $\{x_1, x_2, x_3\}$ might look.

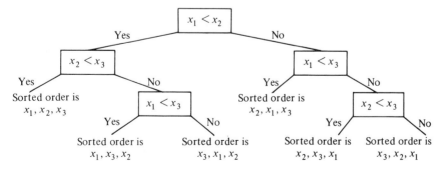

Fig. 2.24 A sorting tree.

Observe that the tree has six leaves, one for each of the 3! = 6 permutations of three keys. This is in general true: for a sorting algorithm to correctly sort n keys, its sorting tree *must* contain at least one leaf for each of the $n!$ permutations on n elements. What is the largest number of leaves in any tree of height h? By a simple induction, that number can be shown to be 2^h; thus for a sorting tree to have $n!$ leaves, it must have at least height h, where $2^h \geq n!$ or $h \geq \log_2 n! = \log_2 (\prod_{i=1}^{n} i) = \sum_{i=1}^{n} \log_2 i \approx n \log_2 n$. But the height of the tree is the length of the longest path from the root to a leaf, and in this case that means the number of comparisons in the worst case. Therefore, in the worst case at least about $n \log_2 n$ comparisons will be required by *any* sorting algorithm.

Proving that the average number of comparisons can never be better than about $n \log_2 n$ is more difficult. We shall not prove it, but the proof is similar to the argument above, except it involves showing that the *average distance* from the root to a leaf in a tree is appropriately bounded, as is the height.

These arguments were based on the comparison of two keys as a unit of work. However, we only used the fact that a sorting algorithm can be expressed in the form of a binary tree flowchart, independently of what tests are performed inside the decision boxes. As long as there are only two possible outcomes to the test, it does not matter whether we use comparisons or some more complex predicates. The proof we have sketched shows that about $n \log_2 n$ binary tests (of any kind) are required on the average to sort n keys.

2.6.   REMARKS AND REFERENCES

Since Leibnitz gave the name to the field, combinatorial mathematics has remained largely dormant until the twentieth century. In recent years a number of textbooks have appeared, for example

> RIORDAN, J. *An Introduction to Combinatorial Analysis*, Wiley, New York, 1958.
>
> RYSER, H. J. *Combinatorial Mathematics*, The Mathematical Association of America, 1963.
>
> BECKENBACH, E. F. (ed.). *Applied Combinatorial Mathematics*, Wiley, New York, 1964.
>
> HALL, M. *Combinatorial Theory*, Xerox College Publishing, Lexington, Mass., 1967.
>
> LIU, C. L. *Introduction to Combinatorial Mathematics*, McGraw-Hill, New York, 1968.

During the 1950s, digital computers began to be applied to problems in this field. Some of these beginnings were discussed in

HALL, M., and D. E. KNUTH. "Combinatorial Analysis and Computers," *Amer. Math. Monthly, 72,* Pt. II (Feb. 1965), 21–28.

Backtrack techniques became common lore of programmers without being attributable to any one person, but they were first described and named in

WALKER, R. J. "An Enumerative Technique for a Class of Combinatorial Problems," in *Combinatorial Analysis (Proceedings of Symposia in Applied Mathematics,* Vol. X), American Mathematical Society, Providence, R. I., 1960.

A further discussion with examples and applications of backtrack can be found in

GOLOMB, S., and L. BAUMERT. "Backtrack Programming," *J. ACM, 12* (1965), 516–524.

The book with the article by Walker also discusses many areas of combinatorial theory, and so gives a good overview of the extent of the field. In the same book, an article by D. H. Lehmer, "Teaching Combinatorial Tricks to a Computer," discusses some problems of programming combinatorial algorithms.

Many of the techniques that have been developed for combinatorial computing are surveyed in

WELLS, M. B. *Elements of Combinatorial Computing,* Pergamon Press, Elmsford, N.Y., 1971.

This book is distinguished by the definition and use of a special programming language written with combinatorial applications in mind. A wealth of information on computer techniques for the efficient implementation of combinatorial algorithms can be found in the excellent book

KNUTH, D. E. *Fundamental Algorithms, The Art of Computer Programming,* Vol. 1, Addison-Wesley, Reading, Mass., 1968.

Block designs are treated in depth in a setting of combinatorial mathematics in Hall's book listed previously; for a discussion of their applications to experimental design, the reader should consult

FINNEY, D. J. *The Theory of Experimental Design,* University of Chicago Press, Chicago, 1960,

or

MANN, H. B. *Analysis and Design of Experiments,* Dover, New York, 1949.

An interesting discussion of combinatorial problems arising in timetable construction can be found in

CSIMA, J., and C. C. GOTLIEB. "Tests on a Computer Method for Constructing School Timetables," *Comm. ACM, 7* (1964), 160–163.

An interesting historical development of the theory of orthogonal latin squares is found in

> STEIN, S. K. *Mathematics—The Man-Made Universe*, 2nd ed., W. H. Freeman, San Francisco, 1969.

Here also are two chapters on tiling with squares and the application of electrical network theory. Problems encountered and some results obtained in a search for orthogonal latin squares are described in

> PARKER, E. T. "Computer Investigation of Orthogonal Latin Squares of Order Ten," *Proceedings of Symposia in Applied Mathematics*, Vol. XV, pp. 73–81, American Mathematical Society, Providence, R. I., 1963.

A fascinating personal account of the research leading to the solution of the perfect-square problem is written by W. Tutte in

> GARDNER, M. *The 2nd Scientific American Book of Mathematical Puzzles and Diversions*, Simon and Schuster, New York, 1961.

Tiling or packing problems seem to be a very popular topic for mathematical puzzles and recreational problems. Various polygons are used for tiles (not just squares); in particular, tiling problems involving polyominoes, figures formed by joining a set of unit squares together, have an enthusiastic following. Frequent papers on this topic are found in

> *Journal of Recreational Mathematics*, Greenwood Press, Westport, Conn.

One example of a backtrack procedure for searching for polyomino tilings, which includes a detailed description of the corresponding computer program, is

> FLETCHER, J. G. "A Program to Solve the Pentomino Problem by the Recursive Use of Macros," *Comm. ACM, 8* (1965), 621–623.

Like combinatorial mathematics in general, graph theory is also largely a product of this century. One of the earliest books in the field is

> KÖNIG, D. *Theorie der endlichen und unendlichen Graphen*, Akademie-Verlag M.B.H., Leipzig, 1936; or Chelsea Publishing Company, New York, 1950.

Some recent textbooks are

> BERGE, C. *The Theory of Graphs and Its Applications*, Wiley, New York, 1962.

> HARARY, F. *Graph Theory*, Addison-Wesley, Reading, Mass., 1969.

> ORE, O. *Theory of Graphs* (Cambridge Colloquium Publications, Vol. 38), American Mathematical Society, Providence, R.I., 1962.

An elementary exposition of graph theory is given in

ORE, O. *Graphs and Their Uses*, Random House and the L. W. Singer Company, New York, 1963.

For a treatment of the applications of graphs to decision theory and dynamic programming see

KAUFMANN, A. *Graphs, Dynamic Programming, and Finite Games* (translated from the French), Academic Press, New York, 1967.

An introduction to the subject of how graph algorithms can be implemented on a computer is given in

READ, R. C. "Teaching Graph Theory to a Computer," in *Recent Progress in Combinatorics*, W. T. Tutte (ed.), Academic Press, New York, 1969, pp. 161–174.

The shortest-path algorithm of Section 2.4.1 was first described in

MOORE, E. F. "The Shortest Path Through a Maze," *Proceedings of the International Symposium on the Theory of Switching*, Harvard University Press, Cambridge, Mass., 1959.

An interesting variation of this algorithm is applicable to graphs that are superimposed on a rectangular grid. It saves memory space by labeling nodes only with 0 and 1 (instead of their distance from a starting node) and is described in

AKERS, S. B. "A Modification of Lee's Path Connection Algorithm," *IEEE Trans. Electronic Computers, 16* (1967), 97–98.

A shortest-path algorithm applicable to graphs with edges of arbitrary length, as well as an algorithm for constructing minimal-cost spanning trees (which is a refinement of the second algorithm in Section 2.4.3), is described in

DIJKSTRA, E. Two Problems in Connexion with Graphs," *Num. Math. 1* (1959), 269–271.

The connectivity algorithm of Section 2.4.2 was first presented in

WARSHALL, S. "A Theorem on Boolean Matrices," *J. ACM, 9* (1962), 11–12.

The subject of just how fast the connectivity matrix of a graph can be computed has received increased attention lately, in connection with the problem of how many operations are required to multiply two matrices. A recent result and further references can be found in

MUNRO, J. I. "Efficient Determination of the Transitive Closure of a Directed Graph, *Information Processing Letters, 1* (1971), 56–58.

The basic idea of Section 2.4.4 used to generate all spanning trees was described in

MINTY, G. J. "A Simple Algorithm for Listing All the Trees of a Graph," *IEEE Trans. Circuit Theory, 12* (1965), 120.

A very compact recursive program based on the same idea was published in

McILROY, M. D. "Generator of Spanning Trees, Algorithm 354," *Comm. ACM, 12* (1969), 511.

The problem of sorting by machine goes back to the late nineteenth century when Herman Hollerith devised machines to process census data. The details of Hollerith's work can be found in

TRUESDELL, L. E. *The Development of Punch Card Tabulation*, U.S. Bureau of the Census, Washington, D.C., 1965.

As programmable computers were developed, sorting problems became the epitomy of combinatorial algorithms, and so considerable effort was spent on them. For an interesting bit of this history see

KNUTH, D. E. "Von Neumann's First Computer Program," *Comp. Surveys, 2* (1970), 247–260.

The literature of sorting is so vast that frequently ideas were published by many different people; in many cases ideas simply became part of the folklore of computers, their origins completely obscured. For these reasons we cannot cite any one reference to give credit for most of the algorithms presented in Section 2.5; for those algorithms whose original source is well known, in each case the original paper is more difficult to read than later textbook accounts. In view of this, it is best to only list secondary references. An easy-to-read survey article is

MARTIN, W. A. Sorting, *Comp. Surveys 3* (1971), 148–174.

A complete bibliography on sorting, listing most papers written on sorting prior to June, 1972, can be found in

RIVEST, R. L., and D. E. KNUTH. "Bibliography 26, Computer Sorting," *Comp. Revs., 13* (1972), 283–289.

Just about everything one might need to know about sorting is nicely and thoroughly discussed in

KNUTH, D. E. *The Art of Computer Programming*, Vol. 3: Sorting and Searching, Addison-Wesley, Reading, Mass., 1973.

2.7. EXERCISES

1. Write an algorithm in the style of Section 2.1 to find all the solutions to the eight-queen problem. Describe the set U and the vector A to be built up by the algorithm.

2. (Instant insanity) Given four cubes with their faces colored as illustrated, find an arrangement of the cubes in a row so that all four colors are displayed

Red	White	Blue
	White	
	Green	
	Red	

	Blue	
	Green	
	Red	
White	Green	Blue

Green	White	Red
	White	
	Green	
	Blue	

	Red	
	White	
	Red	
Red	Blue	Green

in each group of four adjacent faces. In practice it is unlikely that by mere trial-and-error procedures one will stumble upon a correct arrangement (there are $3 \cdot 24^3$ possible). As in Exercise 1, write a backtrack algorithm to search for solutions to this puzzle. Describe in detail the set U and the vector A.

3. Thirty-two moveable pieces are arranged on a board as shown with a blank space in the center. A piece moves on this board by jumping over one of its immediate neighbors (horizontally or vertically) into an empty space opposite. Whenever a piece is so "jumped" by another, it is removed from the board.

<pre>
            x   x   x

            x   x   x

    x   x   x   x   x   x   x

    x   x   x   ○   x   x   x

    x   x   x   x   x   x   x

            x   x   x

            x   x   x
</pre>

The object of the puzzle is to perform a series of jumps so as to remove all pieces but one from the board and leave the remaining piece in the center space. Discuss the organization of data required to solve this puzzle with a backtrack algorithm. That is, what is the set U? How might it be represented in a computer program? How would it be ordered? What steps would be required in the computation of the S_k's?

4. Use a backtrack procedure to prove that there exists no latin square orthogonal to the following:

$$
\begin{array}{cccc}
1 & 2 & 3 & 4 \\
2 & 3 & 4 & 1 \\
3 & 4 & 1 & 2 \\
4 & 1 & 2 & 3 \\
\end{array}
$$

5. Derive a formula by which a pair of orthogonal latin squares of order n can be combined to form a magic square of order n.

6. Prove that it is impossible to find n mutually orthogonal latin squares of order n, for any $n \geq 2$.

7. (a) What electrical network corresponds to a tiling of a rectangle with congruent squares?
   (b) What can you say about the set of tilings obtained from those electrical networks which have the same graph structure but differ only in which nodes are the source and sink?

8. Supply the details for the development of a backtrack procedure to answer the question: Given a set of squares whose combined area equals the area of a given rectangle, is there an arrangement of these squares that will tile the rectangle?

9. Prove by induction that the algorithm for computing the number of distinct shortest paths given in Section 2.4.1 is correct.

10. Devise an algorithm for finding the shortest paths between two given nodes in a graph each of whose edges has been assigned a nonnegative real number as its length. The length of a path is the sum of the lengths of the edges in it.

11. The connectivity matrix C of a graph can be obtained from the adjacency matrix A also by the following process. First, define an operation "$\circ$" on n by n matrices whose entries may assume only the values 0 or 1:

$$C = A \circ B \qquad \text{has entries} \qquad C_{ik} = \bigvee_{j=1}^{n} (A_{ij} \wedge B_{jk}).$$

$\vee$ and $\wedge$ denote logical OR and AND, respectively. Notice that by replacing $\vee$ by summation and $\wedge$ by multiplication this operation would become ordinary matrix multiplication.

   (a) Prove that the operation $\circ$ is associative; that is,

$$(A \circ B) \circ C = A \circ (B \circ C).$$

   (b) Because of associativity, raising a matrix to a power with respect to the operation $\circ$ is a well-defined operation. Prove that the connectivity matrix C of a graph with n nodes is equal to $A \vee I$, the adjacency matrix with 1's on the main diagonal.
   (c) Compute how many operations (as a function of n) are needed to evaluate C in this manner, and compare this to the number of operations required by Warshall's algorithm.

12. The distance matrix D of a graph can be obtained from the edge-length matrix A by taking the $(n-1)$st power of A with respect to another operation, which is formally analogous to matrix multiplication; that is,

$$C = A \circ B \qquad \text{has entries} \qquad C_{ik} = \min_{1 \leq j \leq n} (A_{ij} + B_{jk}).$$

Carry out the same analysis on this operation as was required in Exercise 11.

13. A graph G can be represented by its *fundamental circuit matrix*, which is defined as follows. Choose an arbitrary spanning tree T of G. Each edge e not in T defines a unique circuit with T, consisting of e itself and of the edges in the set we have called $CIR(e, T)$ in Section 2.4.3. These circuits are called "fundamental" (with respect to a given spanning tree T). The fundamental circuit matrix F of a graph G (with respect to T) has a column for each edge of G, and a row for each fundamental circuit. F_{ij} is 1 if edge e_j is in the fundamental circuit c_i, and 0 otherwise. The illustration of a graph with eight edges shows the fundamental circuit matrix with respect to the tree $T = \{a, b, c, d\}$. The

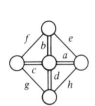

	$\longleftarrow$		T	$\longrightarrow$				
	a	b	c	d	e	f	g	h
e	1	1	0	0	1	0	0	0
f	0	1	1	0	0	1	0	0
g	0	0	1	1	0	0	1	0
h	1	0	0	1	0	0	0	1

fundamental circuits are identified by their unique edge not in T. Notice the identity matrix on the right. Assume an exchange $T' = T - a + e$ is carried out. How can the fundamental circuit matrix F' of G with respect to T' be computed from the fundamental circuit matrix F with respect to T? State this rule in general, and use it to estimate the number of operations required by the exchange algorithm for finding a minimal-cost spanning tree.

14. Prove that algorithms 1 and 2 of Section 2.4.3 construct a minimal-cost spanning tree.

15. Write a program to implement an algorithm for generating all spanning trees based on the idea of Section 2.4.4.

16. Prove that if we change the $<$ branch of a tree flowchart of any sorting algorithm to $\leq$ the algorithm will work, even if keys can be duplicated.

17. Develop and program a nonrecursive version of the Quicksort algorithm. [*Hint:* Use a push-down stack to store ordered pairs (i, j). A pair on the stack means that $x_i, \ldots, x_j$ needs to be sorted by the algorithm. For example, in Figure 2.15 the stack would initially be set to contain the pair $(1, 8)$. After the value $x_5 = 5$ was found to be the point at which the scans met, the pair $(1, 8)$ would be removed from the stack, and ordered pairs $(1, 4)$ and $(6, 8)$ would be put on. The algorithm terminates when the stack is empty.]

18. Prove the theorem stated in Section 2.5.2.

19. Write a program for the tournament-selection sort given in Section 2.5.2.

20. Discuss why it is impossible to use a binary search on a linearly linked list.

21. Develop and program a nonrecursive tree-traversal algorithm (see Section 2.5.4). (*Hint:* Use a push-down stack.)

22. Show that a single application of one of the two transformations pictured in Figure 2.23 (or of its symmetric variant) is sufficient to rebalance any balanced tree which has been disrupted by the insertion of a node.

23. Develop an algorithm for deleting a node from a balanced tree without destroying the balance. (*Hint:* Use the same transformations.)

24. Examine alternative definitions of balance along with corresponding insertion and deletion algorithms. For example, for every node restrict the ratio of the number of nodes in the left subtree to the number of nodes in the right subtree to lie in a fixed interval around 1. In this case, do the same transformations work to rebalance the tree? Does that depend on the interval chosen? Is there any relation between this definition of balance and the one used in Section 2.5.4?

25. Prove that even if we allow *ternary* decision nodes (instead of the binary decision nodes we have discussed) any sorting tree must have height at least about $n \log_2 n$.

3 GAME PLAYING AND DECISION MAKING

At most computer installations there are programs that will play some game against visitors. These programs are usually just considered to have curiosity value, and indeed their existence is mainly due to the fact that many programmers like to play games. There are, however, serious reasons for studying game playing, both from the point of view of developing mathematical models as well as their implementation as computer programs. The primary reason is that games are typical examples of decision-making processes, and an understanding of the principles which are common to all, or at least to a large class of, games may help in making good decisions in real life.

Games are, of course, much simpler than most important decision situations one faces in life. Most games extract from human conflict situations some essential aspects, while ignoring many of the secondary considerations that would inevitably enter into real-life situations. It is precisely this fact that makes them good models for study: one can concentrate on a few aspects at a time. Of course, as in all science, one has to be aware of the difficulties inherent in applying conclusions obtained on simplified models to complex situations.

We shall discuss three aspects of game playing. First, we present three examples of well-known games for which an elegant, complete solution is known, and which make excellent demonstration programs. The solutions are specific to each game, and from these games little can be learned that holds for other games in general. A fourth game, Hex, is described in order to present an idea that is important in many conflict situations: it may be possible to prove that a player has a winning strategy without knowing what this strategy is.

Second, we give a short introduction to a mathematical model (called two-person zero-sum games) of games played between two opponents, in which one player's win is the other's loss, and opponents play according to a *minimax principle*. In principle, this theory is applicable to almost all parlor games that involve two players. In practice, however, in attempting to apply this theory directly one finds that even simple games lead to prohibitively large computations.

Finally, we discuss the techniques that have been developed to make game playing computationally feasible, techniques which bear to game theory the same relation that numerical methods have to classical analysis. These techniques cluster around the notion of tree search, which is of great practical importance not only in game-playing programs, but in many situations where one has to search a set that is too large to be exhausted, and where one has to rely on heuristic shortcuts. The reader will find it instructive to compare this section to the backtrack procedure described in Section 2.1, which is designed for an exhaustive search, in contrast to the techniques described here, which are designed for a selective search along promising lines.

To show what has been achieved with the techniques to be described, let us mention some of the highlights in the short history of game-playing programs. Perhaps the earliest paper on the subject was written by C. E. Shannon in 1950 with the title "Programming a Digital Computer for Playing Chess"; in this paper Shannon also raises the related question of "machine thinking," which we shall discuss in Chapter 6. Ever since this paper, the development of chess-playing programs has continued, and this topic serves as an excellent illustration of how computers have been programmed to perform at a nontrivial level in an activity commonly regarded as having a high intellectual content.

In 1951, A. M. Turing described, but did not actually program, an algorithm for playing chess; its main contribution was the idea of a dead position, or quiescent position, which will be described in Section 3.3.2. By the late 1950s there were several programs, but all of them played chess roughly at the level of a novice. The first program to surpass this level, called Mack Hack VI, was written by R. Greenblatt. It was the first program to play in a regular tournament, and did so with surprising success; in 1967 it won a Class D prize, the next-to-lowest ranked class of the United States Chess Federation. At the annual conference of the Association for Computing Machinery in 1970, and every year since, there were computer chess tournaments, and it appears that several of the participating programs can now play about as well as low-ranked human tournament players. It is remarkable that this steady improvement in playing performance has been achieved without a proportionate increase in program complexity. This is encouraging confirmation that we are beginning to learn important principles of how game-playing programs should be written.

The most successful game-playing program written so far is undoubtedly a checker-playing program, which reached master's level. It was written over a period of 20 years (1947–1967) by Arthur Samuel, who pioneered many of the heuristic game-playing techniques that are discussed later.

Game-playing programs such as these have been an important part of the research effort in machine intelligence, and one may expect that they will continue to provide new tools and techniques which are of general use in heuristic programming.

3.1. SOME GAMES

There are a number of well-known games that pose difficult problems to the uninitiated, but for which a partial or complete solution is known. For one familiar with the analysis, they cease to be games in the usual sense, but it is still possible to enjoy the elegance of the solution. Such games provide convenient demonstration programs, in that a rather simple program can make a computer behave in a way which looks impressive to the unknowledgeable player.

For some games like Hex, which is discussed in Section 3.1.4, a partial solution may prove the existence of a winning strategy, but the nonconstructive nature of the proof leaves the actual playing of the game a challenging task.

3.1.1. Nim

The game Nim is used often as a recreational activity in mathematics at the junior high school level or below. It is an easily played game which has a simple, elegant strategy that can be mastered by anyone who understands the binary representation of integers.

The play of Nim begins with three piles of objects or, equivalently, with three positive integers N_1, N_2, N_3. A player's turn consists of the removal of an arbitrary (positive) number of objects from exactly one of the piles or, equivalently, replacing one of the integers N_i by a new integer N_i' where $0 \leq N_i' < N_i$. The players alternate turns until all the objects are removed, or $N_1 = N_2 = N_3 = 0$; the player who was the last to move is the winner. (In a variation of the game, this player is declared the loser; see Exercise 1.) Depending on the first three numbers, it is possible for either player to force a win. For example, in a game starting from

$$\begin{array}{cc} ||| & 3 \\ |||| & 4 \\ ||||| & 5 \end{array}$$

the first player to move has a winning strategy, while starting from

$$
\begin{array}{ll}
||\ \ \ \ \ \ \ \ & 2 \\
|||||\ \ \ \ & 5 \\
|||||||\ \ & 7
\end{array}
$$

the second player can force a win.

These two triples of initial numbers are representative of the two types of situations that can occur during the play of Nim. Let us say that a triple of numbers is *safe* if the first player has a winning strategy; otherwise, a triple is *unsafe*. The key to Nim is the following characterization of these two types of triples of numbers.

Given a triple (A, B, C), consider the matrix

$$
\begin{array}{ccccc}
a_k & a_{k-1} & \cdots & a_1 & a_0 \\
b_k & b_{k-1} & \cdots & b_1 & b_0 \\
c_k & c_{k-1} & \cdots & c_1 & c_0
\end{array}
$$

where the rows comprise the digits of the binary representations of A, B, and C, respectively, the largest of which is $k + 1$ bits long. Define a function $\mathrm{XOR}(A, B, C)$ by

$$
\begin{aligned}
\mathrm{XOR}(A, B, C) &= (a_k + b_k + c_k, \ldots, a_0 + b_0 + c_0) \text{ modulo } 2 \\
&= (k + 1)\text{-tuple obtained by taking the sum} \\
&\quad \text{modulo 2 of each column of the matrix.}
\end{aligned}
$$

XOR is an abbreviation for "exclusive–or," a frequently encountered function whose arguments and result can assume only the values 0 or 1. For example,

$$
\begin{array}{lll}
0 \ 1 \ 1 & \qquad & 0 \ 1 \ 0 \\
1 \ 0 \ 0 & \qquad & 1 \ 0 \ 1 \\
1 \ 0 \ 1 & \qquad & 1 \ 1 \ 1 \\
\overline{0 \ 1 \ 0} = \mathrm{XOR}(3, 4, 5); & \qquad & \overline{0 \ 0 \ 0} = \mathrm{XOR}(2, 5, 7).
\end{array}
$$

Clearly, $\mathrm{XOR}(A, B, C) = (0, 0, 0)$ only when an even number (0 or 2) of 1's occurs in each column of this matrix.

Consider the value of the XOR function as the plays of Nim are made. First, suppose that

$$
\mathrm{XOR}(A, B, C) = (0, 0, 0)
$$

so that an even number of 1's occurs in each column of the previous matrix; for example,

$$\begin{array}{ccc c} 0 & 1 & 0 & 2 \\ 1 & 0 & 1 & 5 \\ 1 & 1 & 1 & 7 \end{array}$$

Notice that any legal Nim play on such a triple will, since it will affect only one row of the matrix, result in a new triple whose XOR value differs from $(0, 0, 0)$ in at least one component. Thus at least one 1 will have to change to a 0. We have

$$XOR(A, B, C) = (0, 0, 0),$$

followed by any play, yields

$$XOR(A', B', C') \neq (0, 0, 0).$$

Now suppose that $XOR(A, B, C)$ is not equal to $(0, 0, 0)$, as with

$$\begin{array}{ccc c} 0 & 1 & 1 & 3 \\ 1 & 0 & 0 & 4 \\ 1 & 0 & 1 & 5 \end{array}$$

Here, as before, any play will change the value of the XOR function, but in this case it is possible to choose a play that will yield

$$XOR(A', B', C') = (0, 0, 0).$$

This can be done as follows: starting from the left of the matrix, change the columns that contain an odd number of 1's by adding or deleting a 1 to put an even number of 1's in the column. This operation on the matrix can be accomplished by a legal play. (Why?)

The strategy should be clear. If we identify

$$\text{safe} \quad \text{with} \quad XOR(A, B, C) \neq (0, 0, 0)$$

and

$$\text{unsafe} \quad \text{with} \quad XOR(A, B, C) = (0, 0, 0),$$

we see that any player faced with a safe triple can force a return to another safe triple (when it is again his turn to move) by choosing a play that will yield an unsafe triple for his opponent. In particular, the opponent will eventually obtain the unsafe triple (0, 0, 0), which ends the game. Note that either player can take advantage of a safe triple in his turn. If, through ignorance of the strategy or through some error, a player fails to turn a safe triple into an unsafe triple, his opponent can immediately seize control of the situation and force the win.

The important concept to be learned from this example is that of a *property of positions which can be preserved*. In the case of Nim, such a property is the safeness of a position, and it leads to a complete analysis of the game. Properties that can be preserved can be found in many games; a second example is described in the next section.

3.1.2. Shannon Switching Games

Shannon's switching game is played on an arbitrary graph with two distinguished nodes. Two players, called *Cut* and *Short*†, play alternately. On each play, Cut deletes one edge from the graph, while Short claims one edge, which is subsequently immune from deletion by Cut. Short wins if he thus preserves a path that connects the two distinguished nodes; otherwise, Cut

Initial   Short
position   wins

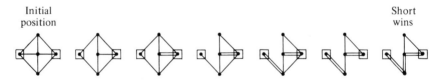

Fig. 3.1 A Shannon switching game in which Cut moves first, but Short wins. Double lines indicate the edges claimed by Short.

wins. A sample game is shown in Figure 3.1. Since no draws are possible, every switching game falls into one of the following three categories:

   1. It is a *Short game* if Short has a strategy that guarantees a win regardless of who plays first.
   2. It is a *Cut game* if Cut has a strategy that guarantees a win regardless of who plays first.
   3. It is a *neutral game* if whoever moves first has a strategy that guarantees a win.

†The name "Short" comes from an analogy between these games and electrical networks.

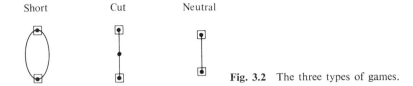

Short   Cut   Neutral

Fig. 3.2 The three types of games.

The simplest examples of each kind of game are shown in Figure 3.2. In these examples it is immediately discernible what kind of game is represented by a simple consideration of all possible moves. On more complex graphs of, say, a dozen edges such a mental analysis becomes practically impossible, and for large graphs (of more than two dozen edges, say), a consideration of all possible moves and countermoves becomes prohibitively long, even at computer speeds. However, it has been shown that one may decide relatively easily on any graph the type of game it represents by using a concept which generalizes the games pictured in the figure. This concept will occupy our attention in what follows.

Consider the graph in Figure 3.3(a). Notice that the set of edges of the graph has been partitioned into two spanning trees T_1 (straight lines) and T_2 (wiggly lines). In such a situation we shall see that no matter what move Cut makes, Short has an answering move which eventually leads to a win for Short. Let us examine this strategy by considering a sample sequence of moves.

As the first move, suppose that Cut removes edge b; the answering move by Short is to preserve one of the edges b_1, b_2, b_3, or b_4. The effect of Short preserving b_1 is shown in Figure 3.3(b). By removing edge b, Cut split the tree T_1 into two disjoint components. Short should reason as follows:

> Any move I make will permanently connect two nodes of the graph resulting in a graph that, for the purposes of the game, is equivalent to a graph in which these two nodes are merged into a single node. Furthermore, the spanning tree that contained the edge which was eliminated by this collapsing, T_2, clearly remains a spanning tree on the new graph. Thus, if I choose an edge that will reestablish the connection between the two components of T_1, the result after my move will be a new graph with one less node but again with a system of two edge-disjoint spanning trees T_1 and T_2.

Now suppose that Cut removes edge c; this splits tree T_2 into two disconnected components, one being the left distinguished node and the other being the remainder of T_2. The answering move by Short must reconnect these two components, so that a proper move is to preserve any one of c_1, c_2, or c_3. One resulting graph, obtained by preserving c_1, and the equivalent collapsed graph are shown in Figure 3.3(c). Obviously, after a finite number of exchanges of this type the resulting graph will be equivalent to a graph containing as a subgraph the graph of Figure 3.3(d), which is clearly a win for Short. We then

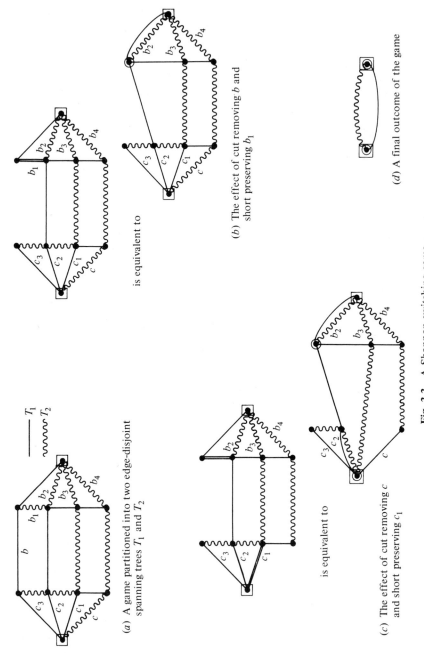

(a) A game partitioned into two edge-disjoint spanning trees T_1 and T_2

— T_1
〜 T_2

is equivalent to

(b) The effect of cut removing b and short preserving b_1

(c) The effect of cut removing c and short preserving c_1

is equivalent to

(d) A final outcome of the game

Fig. 3.3 A Shannon switching game.

104

see that a characterization of a Short game is that the graph contains a system of two edge-disjoint spanning trees. This is the property of a graph that Short can preserve, thus assuring himself a win.

A similar characterization of Cut and neutral games may be made. For example, notice that in the Cut game shown in Figure 3.2 no system of edge-disjoint spanning trees exists, but such a system would exist if two fictitious edges were added directly connecting the two distinguished nodes. These edges are not actually part of the game and are not playable. In this situation there is a strategy by which Cut can foil any attempt by Short to connect the distinguished nodes by any path other than one of the fictitious branches. So, since these are nonplayable, Cut must win the game. Such a strategy might employ the concept of graph duality, which is briefly explained in Exercise 2. The reader is also encouraged to consult the works given in the references for a more thorough discussion.

Finally, to characterize a neutral game, note that in Figure 3.2 a system of edge-disjoint spanning trees can be made by adding only one fictitious edge joining the two distinguished nodes. The reader should convince himself that in such a situation there is always a playable edge whose removal or preservation will result, respectively, in a Cut game or a Short game after one move, so that the first player has a winning strategy.

There is an interesting analogy between Shannon switching games and the algorithm of Section 2.4.4 for generating all spanning trees on a graph. Cut and Short moves correspond exactly to the ways in in which graphs G' and G'' are derived from a graph G, as shown in Figure 2.11.

3.1.3. Calling the Largest Number

Unlike the previous examples in which each of the participants had complete information as to the status of the game at any time, the following example is one in which such information is withheld from one of the players. This is characteristic of most card games, and in such a situation it is appropriate to consider a win or loss of a single game as relatively insignificant and to concentrate on the cumulative history of wins and losses; we are interested in winning or losing "in the long run." In this respect, the game is akin to such games as blackjack or poker.

The game is played as follows. Let one player secretly choose a sequence of n natural numbers of any size. These numbers are recorded and then shuffled, as a deck of n cards would be; then, one at a time, they are revealed to the second player. This player, whom we shall call the *oracle*, looks at each number in turn and has the option of stopping the play at any time by announcing immediately upon seeing a number, "Stop, this is the largest number in the sequence." If the oracle does so, the entire set of numbers is immediately revealed, and it is determined whether the guess was correct. If so, the oracle wins; otherwise, the first player wins.

The important question in this game is how often can the oracle expect to guess correctly, or what odds should he insist upon to at least break even in the long run? One might think that if the number n were very large, the chance that the oracle could guess correctly is small, say near $1/n$, so that the oracle would require long odds in his favor before he would agree to play. The surprising result is that there is a strategy by which the oracle can break even at odds of only 3 to 1 in his favor, regardless of the size of n. For example, if chips were used as a medium of exchange, the oracle would pay one chip when he loses and expect three chips from his opponent when he wins. Before going on, the reader should try to devise a strategy by which the oracle can avoid bankruptcy with these odds.

If you have tried to find your own strategy for playing the game, you are perhaps in a position to appreciate the cleverness of the oracle's strategy. Given the number n beforehand, the oracle always allows an initial fraction f of the numbers to pass without interrupting, and merely remembers the maximum number M that he sees among the numbers in this initial fraction. Thereafter, the oracle picks the first number that is larger than M as the largest. If he never encounters a number larger than M in the rest of the sequence, he remains silent, losing the play.

The pertinent question now is what should the fraction f be? What is the probability, as a function of f, that the oracle will correctly guess the largest number by this method, and what choice of f will maximize this probability? To conveniently estimate this probability $p_n(f)$, we resort to a trick that is often useful in mathematics, but which may puzzle those who have not seen it before. We transform a problem that deals with a finite number of integers into one about the continuum of real numbers. This allows us to replace complicated summations by very simple integrals, greatly simplifying the computational aspects of the solution. The probability $p(f)$ computed in this manner is a good approximation to the probability $p_n(f)$, even for moderate sized n, say when n is 50.

Consider plotting a possible sequence of n numbers as in Figure 3.4(a). In this example, with the maximum number being the xth in the sequence, the oracle's strategy would have succeeded, because in the interval from $f \cdot n$ to x no number occurs that is larger than the maximum in the initial segment from 1 to $f \cdot n$. Now consider that an arbitrary picture of this type has been scaled so that the interval from 1 to n is mapped into the real interval $(0, 1)$ on the x-axis, as shown in Figure 3.4(b). Reinterpreting the problem now on this picture, we see that for the oracle's strategy to work the sequence must have the following properties. First, the largest number must fall somewhere in the

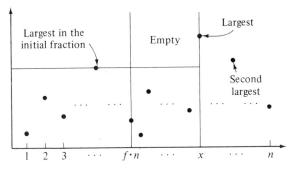

(a) A plot of a possible sequence of n numbers

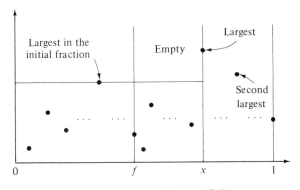

(b) The plot scaled down to (0, 1)

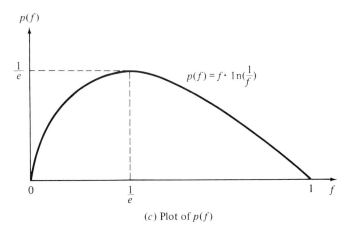

(c) Plot of $p(f)$

Fig. 3.4 The best value of f for the game of calling the largest number.

107

interval $(f, 1)$. If we denote its location by x, one of the following events must occur:

Event	Probability of That Event
Second largest number falls in the interval $(0, f)$	f
or	
Second largest falls in the interval $(x, 1)$ and third largest falls in the interval $(0, f)$	$f \cdot (1 - x)$
or	
Second and third largest fall in $(x, 1)$, and fourth largest falls in $(0, f)$	$f \cdot (1 - x)^2$
or	
Second, third, and fourth largest fall in $(x, 1)$, and fifth largest falls in $(0, f)$	$f \cdot (1 - x)^3$
etc.	

Hence the probability of success, given that the largest number falls at x, is

$$f \cdot (1 + (1 - x) + (1 - x)^2 + (1 - x)^3 + \cdots) = \frac{f}{x}.$$

Since this holds for any x in the interval $(f, 1)$, we find

$$\text{probability of success} = p(f) = f \cdot \int_f^1 \frac{dx}{x} = f \cdot (\ln 1 - \ln f)$$

$$= f \cdot \ln\left(\frac{1}{f}\right).$$

$p(f)$ is plotted in Figure 3.3(c). It has a maximum at $f = 1/e \approx 0.37$ as can be seen by taking the derivative,

$$p'(f) = \ln\left(\frac{1}{f}\right) - 1,$$

and setting it to zero;

$$p'(f) = 0, \quad \Longrightarrow \frac{1}{f} = e \quad \text{or} \quad f = \frac{1}{e}.$$

The probability of success with the fraction $f = 1/e$ is

$$p\left(\frac{1}{e}\right) = \frac{1}{e} \cdot \ln(e) = \frac{1}{e} \approx 0.37.$$

It should be clear why the oracle can take odds of only 3 to 1 in its favor. If the oracle wins the amount Z for every successful identification of the largest number, and loses 1 for every failure, his expected wins are

$$\frac{1}{e} \cdot Z,$$

and his expected losses

$$\left(1 - \frac{1}{e}\right) \cdot 1.$$

Hence if $Z > e - 1 \approx 1.72$, he will win in the long run. One must remember that this analysis holds only in the limit as n approaches infinity, and for a finite sequence the expected win will be slightly different. However, the margin of safety in choosing 3 to 1 odds is such that the oracle can easily win.

3.1.4. Hex

The game of Hex is played on a diamond-shaped board made up of hexagons; the standard-sized board is eleven by eleven, as shown in Figure 3.5. Two opposite sides are labeled "black" and the other two sides are labeled "white"; the hexagons at the corners belong to either side. Correspondingly, there are two players Black and White who move alternatingly, a move consisting of placing a piece (whose color corresponds to that of the player) on a cell not already occupied. The object of the game is for Black to form an

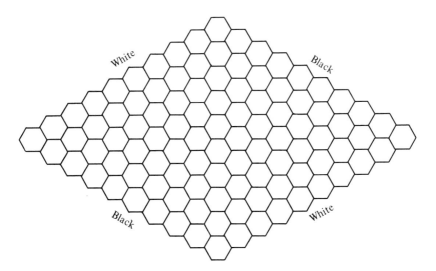

Fig. 3.5   Standard Hex board.

unbroken chain of black pieces between the two sides labeled "black." White's objective is to stop Black, but he can do this *only* by forming a chain of white pieces between the two sides labeled "white"; hence the game cannot end in a draw—someone must win. The chain can have any shape; the only condition is that it connect the opposite sides of the diamond.

Before reading further, the reader should attempt to gain insight into the subtleties of Hex by examining the game on smaller boards. In particular, he should try to prove that on three by three, four by four, and five by five boards the first player can always win.

As the boards become larger, the analysis becomes increasingly more difficult, and the eleven by eleven game has not been fully analyzed. There is, however, a clever proof that on the standard board the first player has a winning strategy. The proof does not exhibit such a strategy; it simply shows that the assumption that such a strategy does not exist leads to a contradiction.

The proof begins with the observation that since draw games are impossible and either the first or second player must win, one of them must, of necessity, have a winning strategy. Suppose that the second player has a winning strategy. The first player can then play as follows: he makes an arbitrary move as his first move and then, pretending he is the second player, he assumes the second player's winning strategy. If in following this strategy he is required to play on the position occupied by his arbitrary move, he

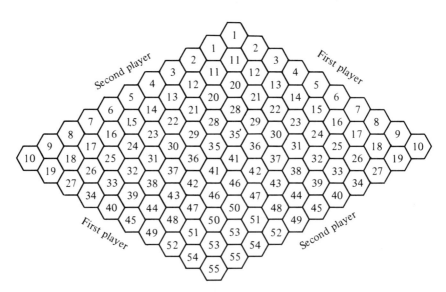

Fig. 3.6 The second player can always win on this eleven by ten board by "imitating" the first player and playing on the position with the same number as the first player's move.

makes, instead, another arbitrary move. He continues playing in this fashion, always having an extra piece on the board. Now, this extra piece can in no way interfere with the first player's imitation of the winning strategy, since the extra piece is clearly not a liability. So, the first player can win, contradicting the assumption that the second player has a winning strategy. Thus it must be the first player who has the winning strategy.

This rather elegant proof implicitly uses some properties of the standard board. If we were playing on an eleven by ten board instead, the *second* player could always win, provided the first player must connect the sides that are farther apart. The second player's strategy is to always play in the cell labeled with the same number as his opponent's previous move, as shown in Figure 3.6. We leave the explanation of why the proof does not apply in this case for the reader to discover.

3.2. BASIC IDEAS OF GAME THEORY

Games have inspired achievements that subsequently turned out to have a significance beyond the original expectations. One example of this is found in the questions concerning the odds in dice games which a seventeenth century French gambler, Chevalier de la Meré, asked the famous mathematician Pascal. Pascal went on to found the theory of probability, which today is one of the indispensable tools of science.

More recently, in 1928, von Neumann wrote a paper "On the Theory of Parlor Games," which was the foundation of the mathematical theory of games. It was soon recognized that game theory is not nearly as relevant to playing games (we shall discuss the principal techniques used in game-playing programs in Section 3.3) as it is for modeling situations of decision and conflict in economic activities. Indeed, the next major publication in game theory, in 1944, was a book by von Neumann and Morgenstern entitled *Theory of Games and Economic Behavior.*

Since then game theory has grown into a diverse subject, which straddles the borderline between mathematics, operations research, and theoretical economics. It is too early to say whether it will ever gain an importance comparable to that of probability theory, but one thing is clear: game theory was the first mathematical attempt to capture the notion of decision making in conflict situations. One can study how to optimize a quantity, one's "payoff," where one cannot manipulate the variables freely (as is possible in the classical optimization problems of the calculus), and where one cannot assume that these variables take on values at random according to a known distribution (as in the problem of optimizing expected values in probability theory), but where one must assume the existence of an intelligent opponent who controls some of the variables and who will do his utmost to spoil the optimization attempt. This entirely novel aspect of game theory may hold the key to important applications in economic and social sciences, and therefore we include a discussion of a small part of it.

3.2.1. Two-Person Zero-Sum Games

A two-person zero-sum game is played on an m by n *payoff matrix A* between two opponents, P and Q, as follows. P selects an integer i, $1 \le i \le m$, and Q selects an integer j, $1 \le j \le n$, each without having any knowlege of the other's choice. Thereupon both choices are revealed, and Q pays to P the amount A_{ij}, the entry in row i and column j of the payoff matrix. If A_{ij} is negative, this amounts to P paying to Q the amount $|A_{ij}|$.

This is called a zero-sum game because one player's win is the other's loss, so that the sum of their fortunes is the same before and after the game has been played, P's fortune having changed by $+A_{ij}$, Q's by $-A_{ij}$. At first this game may seem dull, and more important, of a highly special nature, but we dispel the first of these objections with the next example, and the second one by showing how most parlor games played between two opponents fit this model.

Look at the game

$$Q$$

		1	2	3	4	min of row
	1	1	3	2	9	1
P	2	5	4	4	6	4
	3	8	2	6	2	2
max of column		8	4	6	9	4 / 4

This game is clearly favorable to P, but just *how favorable* is it? How much money would one have to offer to Q, as a reward, so that he might agree to play this game? These questions aim at abstracting a quantity called the "value of the game," which measures how much P can expect to win and how much Q must expect to lose (or vice versa) by playing this game. The phrase "can expect to win or lose" needs a definition, which we shall now develop.

Consider the following two questions:

1. What is the largest quantity L which P can guarantee that he will win?
2. What is the least quantity U which Q can guarantee that he will lose no more than?

We can best answer question 1 by following P's reasoning, which might go as follows: "If I play $i = 1$, the first row, I may win as little as 1," and he records this quantity in an additional column. "If I play $i = 2$, I may win as little as 4, and if I play $i = 3$, as little as 2. Hence the largest quantity I can be sure to win is the maximum of these minima, or 4." This argument yields

$$L = \max_i \min_j A_{ij}.$$

It is clear that since P can guarantee to win at least L the game must be worth at least that much to him, and hence that he will require that the (yet to be defined) value V of a game satisfy the inequality

$$L \leq V.$$

We can find the quantity U similarly by letting Q ask, "What is the largest loss that I might suffer by playing a given column?", and then minimizing over these column maxima to obtain

$$U = \min_i \max_j A_{ij},$$

which in our example also happens to be 4. And again, in analogy with the previous case, our interpretation of the value of the game forces us to insist that

$$V \leq U.$$

From these two inequalities it follows that

$$L \leq V \leq U.$$

For this reasoning to be consistent, it is necessary that $L \leq U$ for any matrix A, and the quantities L and U be defined as above. The reader is encouraged to prove this.

Whenever $L = U$, these inequalities define the value V of the game. This is the case in our example, for which we have $V = 4$. How can one characterize matrices with $L = U$ that is, for which

$$\max_i \min_j A_{ij} = \min_i \max_j A_{ij}?$$

In our example we see that the common value $L = U = 4$ arises from the element $A_{22} = 4$, which has the following properties: it is minimal in its row and maximal in its column. A position (k, l) in a matrix A is called a *saddle point* if it has these two properties:

$$A_{kl} = \min_j A_{kj} \quad \text{and} \quad A_{kl} = \max_i A_{il}.$$

The reader should have no great difficulty in proving the following two
theorems:

1. For any matrix A, $L = U$ if and only if A has a saddle point.
2. For any matrix A, if A has several saddle points, they all have the same
value.

Let us call a player who reasons as in our example, that is, one who
optimizes the outcome he can be guaranteed of obtaining, a *minimax player*.
On a game with a saddle point, two minimax players will inevitably choose a
row and column that intersect on a saddle point.

Minimax play is optimal in the following sense: on a game with a saddle
point (and, as we shall see later, on any game if the notions are appropriately
generalized), a player cannot expect to gain anything by deviating from a
minimax strategy against a minimax player. Yet it must be emphasized that
minimax play is only one of various reasonable philosophies of playing; in
certain circumstances, completely different strategies are more rational than
minimax play. If in our example it were a matter of life or death for P to win
9 units, and no smaller amount would do him any good, then he should
clearly gamble and play row 1, even though by doing so he will probably gain
only 3 units instead of the 4 he could have had for certain.

Not all matrices have saddle points; consider

$$\begin{pmatrix} 0 & 1 \\ 2 & 0 \end{pmatrix}$$

for which $L = 0$ and $U = 1$. If a game without a saddle point is to be played
many times between the same opponents, whether or not they are minimax
players, the strategies chosen by either player will not settle on any single row
or column, as in the case of minimax players playing a saddle-point game,
but will oscillate between various rows and columns. In the previous matrix
this can be seen as follows. P might begin by playing the second row several
times, because, a priori, it looks better than the first row. However, Q will
then quickly realize that he ought to play the second column in order to avoid
loss. After a few more games, P will catch on to this situation and switch to
the first row, thus hoping to win 1 unit. It will then be Q's turn to realize this
change in strategy on P's part, and to respond to it by switching to the first
column. In due time this will cause P to change to the second row, thus
bringing us back to the starting situation.

This discussion indicates that we need a new notion of strategy which
reflects the fact that, if a game is played repeatedly, a player may not (and in
games without saddle point, should not) commit himself to any single row or
column, but instead may want to choose different rows or columns with

certain relative frequencies. In a single game, he would assign a probability to each of his rows or columns equal to the relative frequency with which he wants to play this row in the long run, and then make his choice with the help of a random device that implements the probabilities he has chosen.

To emphasize the distinction between our original notion of strategy (choosing a row or column) and the present one (assigning probabilities to each row or to each column), we speak of a pure strategy in the first case and of a mixed strategy in the second. The latter notion is clearly a generalization of the first, since we can identify a pure strategy with a mixed strategy where one row is assigned probability 1 and all others 0. From now on, we shall use "strategy" to mean "mixed strategy," but remember that this includes "pure strategy" as a special case.

A strategy for P now is a probability vector $\mathbf{p} = (p_1, p_2, \ldots, p_m)$, where $p_i \geq 0$ and $\sum_{i=1}^m p_i = 1$. Similarly, a strategy for Q is a vector $\mathbf{q} = (q_1, q_2, \ldots, q_n)$. The expected outcome of the game, as a function of $\mathbf{p}$ and $\mathbf{q}$, is $E(\mathbf{p}, \mathbf{q}) = \sum_{i=1}^m \sum_{j=1}^n p_i q_j A_{ij}$. It is this value which replaces the quantity A_{ij} in our preceding analysis, that is, the one which P attempts to maximize by appropriate choice of $\mathbf{p}$, and which Q wants to minimize by appropriate choice of $\mathbf{q}$.

As before, if P and Q are minimax players, P is interested in the greatest number L that he can guarantee to win, but now in the sense of the expected payoff $E(\mathbf{p}, \mathbf{q})$, and Q is interested in the least number U which he can guarantee that he will not lose more than, again in the sense of the expected payoff.

$$L = \max_{\mathbf{p}} \min_{\mathbf{q}} E(\mathbf{p}, \mathbf{q}),$$

$$U = \min_{\mathbf{q}} \max_{\mathbf{p}} E(\mathbf{p}, \mathbf{q}).$$

Again, as before, we derive the inequalities

$$L \leq V \leq U,$$

which express the requirement that the value of the game, no matter how it is defined in detail, must lie between these two bounds.

It turns out that we have actually defined the value of a game, since in the context of mixed strategies L is necessarily equal to U. This is expressed by von Neumann's *minimax theorem*: for an m by n matrix A and for all probability vectors $\mathbf{p} = (p_1, p_2, \ldots, p_m)$ and $\mathbf{q} = (q_1, q_2, \ldots, q_n)$, we have

$$\max_{\mathbf{p}} \min_{\mathbf{q}} \sum_{i=1}^m \sum_{j=1}^n p_i q_j A_{ij} = \min_{\mathbf{q}} \max_{\mathbf{p}} \sum_{i=1}^m \sum_{j=1}^n p_i q_j A_{ij}.$$

The quantity thus defined is called the *value* V of the two-person zero-sum game represented by the matrix A.

An optimal minimax strategy for player P is a probability vector $\mathbf{p}_{opt}$ such that

$$\min_{\mathbf{q}} E(\mathbf{p}_{opt}, \mathbf{q}) \geq V,$$

and an optimal strategy for Q is a probability vector $\mathbf{q}_{opt}$ such that

$$\max_{\mathbf{p}} E(\mathbf{p}, \mathbf{q}_{opt}) \leq V.$$

It is a direct consequence of the minimax theorem that optimal strategies exist. They need not be unique, however; for example, on a payoff matrix with all entries zero, any strategy whatsoever is clearly optimal.

As an illustration of these ideas, let us find the value and optimal strategies of the game with payoff matrix

$$\begin{pmatrix} 0 & 1 \\ 2 & 0 \end{pmatrix}.$$

Let $\mathbf{p} = (p, 1 - p)$ and $\mathbf{q} = (q, 1 - q)$ so that

$$E(\mathbf{p}, \mathbf{q}) = p \cdot q \cdot 0 + p \cdot (1 - q) \cdot 1 + (1 - p) \cdot q \cdot 2 + (1 - p) \cdot (1 - q) \cdot 0$$
$$= p - 3 \cdot p \cdot q + 2 \cdot q.$$

Figure 3.7(a) shows the contour lines of $E(\mathbf{p}, \mathbf{q})$. Notice that a saddle point (which was not present in the matrix A) appears at $p = \frac{2}{3}$, $q = \frac{1}{3}$, with value $E(p, q) = \frac{2}{3}$. Notice also how the optimal strategy $p = \frac{2}{3}$ for player P shows up as a contour line that is orthogonal to the p-axis. This means that, regardless of Q's strategy, playing the first row with probability $\frac{2}{3}$ assures P an expected payoff of $\frac{2}{3}$. Analogous arguments hold for Q's optimal strategy, $q = \frac{1}{3}$.

It appears perhaps surprising that the optimal minimax strategy requires P to play the first row twice as frequently as the second row, even though the latter "looks better." Figure 3.7(a) shows why this is so; if P chose any other strategy (say $p = \frac{1}{3}$, playing the second row twice as frequently as the first), he might hope for a higher expected payoff ($\frac{4}{3}$), but he also risks a lower expected payoff ($\frac{1}{3}$), depending on Q's strategy.

Figure 3.7(b) shows an alternative diagram for $E(\mathbf{p}, \mathbf{q})$, where E is plotted as a function of q for various values of p. Because $E(\mathbf{p}, \mathbf{q})$ is a linear function of q, this diagram is easier to plot, but the optimality of the strategies $p = \frac{2}{3}$, $q = \frac{1}{3}$ and the value $V = \frac{2}{3}$ of the game can be seen just as easily.

How does this type of game serve as a model for parlor games in which players typically alternate making moves? The correspondence is derived from the fact that a strategy is to be thought of as a set of rules which completely determines every move that is to be played throughout the game. If

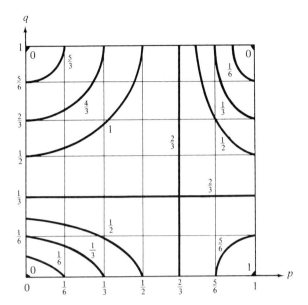

(a)   Contour lines of $E(\mathbf{p}, \mathbf{q}) = p - 3pq + 2q$. The number written
by each curve is its height; for example $E(\frac{1}{3}, \frac{1}{6}) = \frac{1}{2}$,
so the point $(\frac{1}{3}, \frac{1}{6})$ is on a contour line whose height is $\frac{1}{2}$.

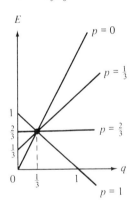

(b)   Another representation of $E(\mathbf{p}, \mathbf{q})$

Fig. 3.7 Strategies for a game with payoff matrix $\begin{pmatrix} 0 & 1 \\ 2 & 0 \end{pmatrix}$.

you had written up your strategy for playing chess or bridge, say, you would
not need to ever be physically present at a game; you could simply give your
strategy to someone, say a computer, who would play for you just the same
way you would have. However, it is clear that for most intellectually challenging games it would be extremely difficult to write down a strategy that plays a

reasonably good game; if this were not the case, games like chess would cease to be played as games. Even if somebody worked very hard writing down such a strategy for a game in which he is an expert, he must expect that his strategy will be poorer than he is, for it is unlikely that he will be able to formulate all the principles that he uses to evaluate positions. This observation is supported by the considerable amount of experience that has been gained in developing game-playing programs. Writing a game-playing program means writing down a strategy for a game. We know of only one well-documented case in which such a strategy has turned out to be much better at the game than its designer—Samuel's checker-playing program mentioned in the introduction.

3.2.2. Fictitious Play for Approximating the Value of a Game

In the preceding section we solved a two-person zero-sum game, finding its value and the optimal strategies for both players, using a graphical technique; this had the advantage that each curve drawn had an intuitively meaningful interpretation. However, it is clear that for payoff matrices which are only moderately larger than the 2 by 2 case treated, graphical techniques become impractical, since each additional row or column introduces a new independent variable.

A systematic computational technique for solving two-person zero-sum games is obtained by associating a linear programming problem with a game, so that the solution of that problem gives the value of the game and the optimal strategies for each player. This associated problem can be solved by any of the standard techniques of linear programming. We shall not pursue this further, since we are interested primarily in the concepts that game theory provides, rather than in computational techniques.

Let us, however, describe one technique for solving two-person zero-sum games, because of the insight it provides into the nature of optimal strategies. It is an iterative technique which generates a sequence of numbers that converges to the value of the game. It also generates sequences from which the probabilities of the optimal strategies can be deduced, but these sequences need not converge unless the optimal strategies are unique. The convergence is slow, so this is not an efficient computational technique; but it may be useful if one only wants a rough estimate of the value of a game, since the method provides a lower and upper bound for the value at each step of the iteration.

This algorithm is due to G. W. Brown who called it the *method of fictitious play*. This descriptive name reflects the fact that this method simulates two opponents who play the same two-person zero-sum game repeatedly. At every step of the iteration, each of the two fictitious players makes use of all the experience he has gathered in the preceding steps. The guiding idea behind this method was expressed by Brown as follows:

> ... The iterative method in question can be loosely characterized by the fact that it rests on the traditional statistician's philosophy of basing future decisions on the relevant past history. Visualize two statisticians, perhaps ignorant of minimax theory, playing many plays of the same discrete zero-sum game. One might naturally expect a statistician to keep track of the opponent's past plays and, in the absence of a more sophisticated calculation, perhaps to choose at each play the optimal pure strategy against the mixture represented by all the opponent's past plays.

Let us introduce the following notation to describe fictitious play. At times $t = 1, 2, 3, \ldots$, two opponents, called P and Q, play a two-person zero-sum game defined by the payoff matrix A. Denote by i_t, j_t the pure strategies (row and column, respectively) chosen by the players P, Q at time t, and denote by $\mathbf{p}_t$, $\mathbf{q}_t$ the mixed strategies that correspond to the frequency with which P and Q, respectively, have played their pure strategies up to time t. To illustrate this, consider the example of the payoff matrix

$$
Q
$$
$$
P \quad \begin{pmatrix} 0 & 1 \\ 2 & 0 \end{pmatrix}
$$

discussed in the preceding section. If, up to time $t = 3$, P has played the rows $i_1 = 2$, $i_2 = 1$, $i_3 = 1$, then his mixed strategy $\mathbf{p}_3$ would be the probability vector $(\frac{2}{3}, \frac{1}{3})$, reflecting the fact that he played the first row two out of three times, and the second row once. Finally, denote by

$$
E(u, \mathbf{q}_t) = \frac{1}{t} \sum_{k=1}^{t} A_{u, j_k}
$$

the expected payoff of playing pure strategy u against mixed strategy $\mathbf{q}_t$, and by

$$
E(\mathbf{p}_t, v) = \frac{1}{t} \sum_{k=1}^{t} A_{i_k, v}
$$

the expected payoff of playing mixed strategy $\mathbf{p}_t$ against pure strategy v.

The fictitious play proceeds as follows:

1. Arbitrarily choose i_1, j_1.
2. For $t \geq 1$, choose i_{t+1} as any u that maximizes $E(u, \mathbf{q}_t)$, and j_{t+1} as any v that minimizes $E(\mathbf{p}_t, v)$.

The table contains the calculations for the first 21 steps of fictitious play for the game with payoff matrix

$$\begin{pmatrix} 0 & 1 \\ 2 & 0 \end{pmatrix}.$$

Instead of recording the quantities $E(u, \mathbf{q}_t)$ and $E(\mathbf{p}_t, v)$ the table instead contains these quantites multiplied by t, since the latter satisfy the following convenient recurrence relation:

$$(t + 1)E(u, \mathbf{q}_{t+1}) = tE(u, \mathbf{q}_t) + A_{u, j_t},$$
$$(t + 1)E(\mathbf{p}_{t+1}, v) = tE(\mathbf{p}_t, v) + A_{i_t, v}.$$

t	i_t	j_t	$t \cdot E(u, \mathbf{q}_t)$		$t \cdot E(\mathbf{p}_t, v)$		U_t	L_t
			$u = 1$	$u = 2$	$v = 1$	$v = 2$		
1	2	2	1	0	2	0	$\to 1$	$0 \leftarrow$
2	1	2	2	0	2	1	1	$\frac{1}{2} \leftarrow$
3	1	2	3	0	2	2	1	$\frac{2}{3} \leftarrow$
4	1	1	3	2	2	3	$\to \frac{3}{4}$	$\frac{1}{2}$
5	1	1	3	4	2	4	$\frac{4}{5}$	$\frac{2}{3}$
6	2	1	3	6	4	4	1	$\frac{3}{4}$
7	2	2	4	6	6	4	$\frac{6}{7}$	$\frac{4}{7}$
8	2	2	5	6	8	4	$\frac{3}{4}$	$\frac{1}{2}$
9	2	2	6	6	10	4	$\to \frac{2}{3}$	$\frac{4}{9}$
10	1	2	7	6	10	5	$\frac{7}{10}$	$\frac{1}{2}$
11	1	2	8	6	10	6	$\frac{8}{11}$	$\frac{6}{11}$
12	1	2	9	6	10	7	$\frac{3}{4}$	$\frac{7}{12}$
13	1	2	10	6	10	8	$\frac{10}{13}$	$\frac{8}{13}$
14	1	2	11	6	10	9	$\frac{11}{14}$	$\frac{9}{14}$
15	1	2	12	6	10	10	$\frac{4}{5}$	$\frac{2}{3}$
16	1	1	12	8	10	11	$\frac{3}{4}$	$\frac{5}{8}$
17	1	1	12	10	10	12	$\frac{12}{17}$	$\frac{10}{17}$
18	1	1	12	12	10	13	$\frac{2}{3}$	$\frac{5}{9}$
19	2	1	12	14	12	13	$\frac{14}{19}$	$\frac{12}{19}$
20	2	1	12	16	14	13	$\frac{4}{5}$	$\frac{13}{20}$
21	2	2	13	16	16	13	$\frac{16}{21}$	$\frac{13}{21}$

Thus each vector in the two columns labeled $t \cdot E(u, \mathbf{q}_t)$ can be computed from the preceding one by adding the column of the payoff matrix determined by j_t, and each vector in the two columns labeled $t \cdot E(\mathbf{p}_t, v)$ can similarly be computed by adding the row determined by i_t.

The table also contains the quantities

$$U_t = \max_u E(u, \mathbf{q}_t) \quad \text{and} \quad L_t = \min_v E(\mathbf{p}_t, v),$$

which provide upper and lower bounds for the value of the game. The arrows in the two adjacent columns emphasize bounds that are sharper than the ones obtained earlier. Notice that in this example we have the exact value of $\frac{2}{3}$ after nine iterations, because the lower bound L_3 and the upper bound U_9 are both $\frac{2}{3}$.

Notice that the change of pure strategies chosen by each player follows the pattern mentioned in our discussion of this game in Section 3.2.1. After the arbitrary start $i_1 = j_1 = 2$, P realizes he should switch to the first row; then Q changes to the first column to avoid loss of 1 unit, P switches back to the second row to gain 2 units, whereupon Q will go back to the second column. The table shows two complete cycles of this kind.

The probabilities in the optimal strategies are approximated by the relative frequencies with which players choose their rows and columns, respectively. The analysis of this game in Section 3.2.1 showed that the optimal strategy for P is to play the first row $\frac{2}{3}$ of the time. In our fictitious play, P played that row 13 out of 21 times. Similarly, Q played the first column 8 out of 21 times, which is a good approximation to the probability $\frac{1}{3}$ of his optimal strategy.

Fictitious play shows how an optimal strategy can be considered to be a mixture of pure strategies, each of which is optimal against the mixture of the opponent's past plays. Moreover, it shows how an optimal strategy can be learned by trial and error.

3.3. GAME TREES AND THEIR EVALUATION

The play of most games unfolds in a series of more-or-less well-defined moves, such as a move of a certain chess piece, the deal of a deck of cards, a bet, a call, and so forth. At any given time during the play of a game the alternative moves available to a player are governed by the playing rules and depend strongly on the history of the previous moves of the game. The totality of the alternative moves at any point and the relationship between these moves and the previous history can be represented schematically in the form of a tree in the following way. Starting from a single distinguished node, called the root, the tree is constructed by attaching one branch that corresponds to each possible first move, and from the end points of these branches, in turn, attaching branches that correspond to the set of answering moves available to the next player. The construction of the tree continues in this manner, adding branches that correspond to the possible answering moves until they are exhausted, when the rules of the game signal the end of play. For the purposes of the algorithms in this section, game trees will be considered *ordered trees* (the branches emanating from a given node will be ordered from left to right).

The nodes of a game tree are divided into levels; the nodes at the kth level correspond to the possible positions that can be reached after k moves in the game. As a special case, the root has level 0. In a game with two players who move alternatingly, these levels are partitioned into two classes, even and odd, depending on whether it is the first or the second player's turn to move. The generalization to n players and to more complicated rules for determining whose turn it is to move is possible, but we shall assume two-person games with alternating moves.

In theory, a game tree can be constructed for any game, but in practice the number of branches in the tree makes this quite unmanageable. An example of the game tree for a game of Nim starting with (1, 2, 3) appears in Figure 3.8. Even for this trivial game the tree is quite large, and for a more complicated game in which the choice of moves is large (say, chess or contract bridge) the tree, though finite, is so large that it is impossible to construct it in its entirety. Nevertheless, the concept of a game tree is a useful one, and much of what game-playing programs do is best described as processing a small portion of this tree.

One characteristic of a skillful game player is the ability to look ahead in the game and choose the most advantageous next move. In light of the above discussion, this "look ahead" is precisely a search along the branches of the game tree where the nodes of the tree are evaluated in terms of payoff values for the player. In the next section we shall discuss a technique for the evaluation of game trees, whereby the evaluation of all the nodes of the tree can be made, given values for all the terminal nodes, that is, the payoff values for the possible outcomes of the game.

3.3.1. Minimax Evaluation and Alpha–Beta Pruning

The best-known technique for evaluating game trees is known as *minimax evaluation*. Suppose that we are concerned with a two-person game so that the levels in the game tree are partitioned into classes, even and odd. Furthermore, let the terminal nodes of the tree, which represent ending conditions of the game, be assigned the value of the payoff to the player who moves first. Then, if we assume that the first player seeks to maximize his payoff and his opponent to minimize his loss, the value of the game for the first player can be determined by doing the following step as long as possible:

To any node N, all of whose successor nodes $N_1, N_2, \ldots, N_k$ have been assigned values $V(N_1), V(N_2), \ldots, V(N_k)$, assign the value

$$V(N) = \begin{cases} \max\,(V(N_1), \ldots, V(N_k)) & \text{if the level of } N \text{ is even,} \\ \min\,(V(N_1), \ldots, V(N_k)) & \text{if the level of } N \text{ is odd.} \end{cases}$$

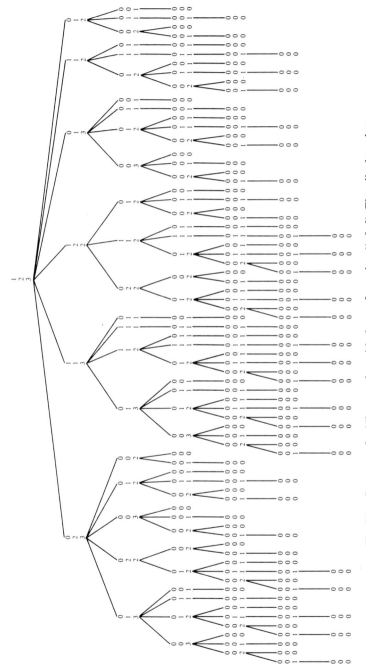

Fig. 3.8 Game tree for Nim, starting with the configuration (1, 2, 3). The piles have been arranged in increasing order; for example, (1, 0, 2) and (1, 2, 0) would both be written as (0, 1, 2).

123

If we assume that the players use a minimax strategy, the algorithm causes each nonterminal node of the tree to be labeled with the payoff that would be expected if the play continued from that point to the end. In particular, therefore, the root of the tree is labeled with the value of the game for the first player. For example, look at Figure 3.9; here the terminal nodes are labeled 0 for a win for the first player in the maximum number of moves, -1 for a win by the second player, and 1 for an "early" win by the first player. The small number written next to each node is the value assigned to that node by the minimax-evaluation algorithm. Note that the root has label 0, indicating that the first player can force a win in the maximum number of moves, the same result established in Section 3.1.2. We leave as an exercise the evaluation of the larger tree of Figure 3.8. (What value should the root of this tree have?)

Since minimax evaluation of trees can be a very lengthy procedure, considerable work has been directed toward improving its efficiency, and with some success. One technique that is quite effective is called alpha–beta pruning. It yields exactly the same value for the root of the tree as the general minimax algorithm, but it saves considerable effort by avoiding those branches whose evaluation would not affect the ultimate value of the root. The circumstance that allows this saving of effort can be observed in the example shown in Figure 3.10. We begin a minimax evaluation using the given values for the terminal nodes. The previous algorithm, as stated, says nothing about the order in which to evaluate the unlabeled nodes, but in any practical implementation some order must be established. So, let us say that among the nodes to which the step of the algorithm applies (i.e., those unlabeled nodes whose immediate successors have been labeled) the leftmost will be evaluated first. For the example in Figure 3.10 this means alphabetical order according to the letter labeling of the nodes. The alpha-beta technique proceeds exactly as a minimax evaluation, but at the same time accumulating as much information as possible about the tree as nodes are evaluated. Note that as soon as a node is evaluated, something is known about the value which will be assigned to its immediate predecessor. In particular, if its predecessor is on a max level in the tree, then the value assigned to the node is a lower bound on the value of its predecessor. Such a lower bound at a max level is called an α-value. In the symmetrical situation where the predecessor node lies on a min level, the resultant upper bound on its value is called a β-value.

In the example, as soon as node d is assigned its value of 3, we immediately have a β-value of 3 for node m, so we know that the value at m will not be larger than 3. With this information we reason as follows: if the value of any other descendant of node m can be shown to be greater than 3, we may safely ignore the rest of the tree below that descendant; thus we can then act as if those branches and nodes had been pruned from the tree. Such a situation is termed a β-cutoff, and it occurs whenever a node two levels below a

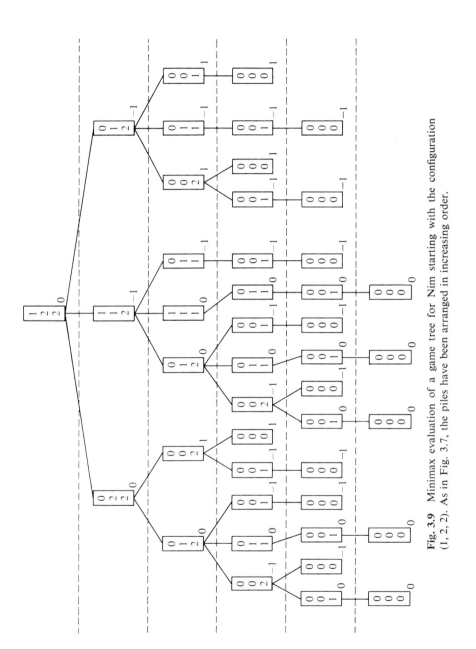

Fig. 3.9 Minimax evaluation of a game tree for Nim starting with the configuration (1, 2, 2). As in Fig. 3.7, the piles have been arranged in increasing order.

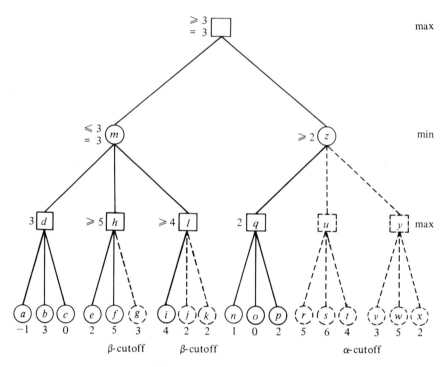

Fig. 3.10 Alpha–beta pruning.

node with a β-value is found to have a value larger than that β. The situation at a max level is the same. In the example, again, having found the value 3 for node m, we have an α of 3 for the root of the tree, so that if node z can be shown to have a value less than 3, we can ignore anything below it. Thus after establishing the value of 2 at node q, we cut off the rest of the tree below node z, since its evaluation will no longer influence the ultimate value of the root. This is called an α-cutoff.

The advantage of this procedure over the exhaustive minimax algorithm is obvious, since with it we need to check only a fraction of the nodes of the tree instead of all of them. We immediately notice, however, that the amount of work saved is greatly influenced by the ordering of the nodes in the tree, since this determines the time at which α- and β-cutoffs occur. Despite this obvious dependence on the order of the search, we can say something about the relative efficiency of alpha–beta pruning over the straight forward minimax algorithm. Consider the tree pictured in Figure 3.11. Here the value A is a lower bound for the value of the root C. Assuming, as above, that evaluation of the nodes B_i takes place from left to right, from B_1 toward B_n, we may define d, the "distance to α-cutoff," as the least $i \leq n$, such that $B_i \leq A$. We want to see how the quantity d behaves as n varies, and in order to do so we

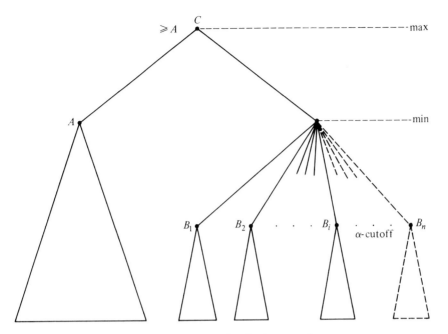

Fig. 3.11 Analysis of alpha–beta pruning.

make some assumptions about the values for the nodes. First, we know that among the values B_i some fraction p of them will be greater than A, and $(1 - p)$ of them less than or equal to A. We assume more generally that there is a p, independent of n, such that each B_i has probability p of being greater than A, independently of any of the other values B_j. Second, we assume that $B_1, \ldots, B_n$ are randomly ordered with respect to their size; that is, they do not occur, say, in order of increasing or decreasing size, but randomly. Then we define the "average distance to α-cutoff":

$$D(n, p) = \sum_{k=1}^{n} k \cdot \Pr(\text{cutoff at } k)$$

$$= \sum_{k=1}^{n} k \cdot \Pr(B_1 > A, B_2 > A, \ldots, B_{k-1} > A, \text{ and } B_k \leq A)$$

$$= \sum_{k=1}^{n} k \cdot p^{k-1} \cdot (1 - p)$$

$$= (1 - p) \sum_{k=1}^{n} k \cdot p^{k-1}$$

where $\Pr(x)$ denotes the probability of event x.

The behavior of this quantity as n increases is clear. For $p < 1$ we have partial sums of a convergent series, so that $D(n, p)$ is always bounded above by its limit, $D(\infty, p)$. This says that, on a tree satisfying the above assumptions, alpha–beta pruning has the effect of limiting the branching of the tree to some value which is independent of the actual branching of the tree. In fact, using the formula

$$\sum_{k=1}^{\infty} p^k = \frac{p}{1-p} \qquad \text{for } p < 1,$$

we compute

$$
\begin{aligned}
D(\infty, p) &= (1-p)\cdot(1 + 2\cdot p + 3\cdot p^2 + 4\cdot p^3 + 5\cdot p^4 + \cdots) \\
&= (1-p)\frac{d}{dp}(p + p^2 + p^3 + p^4 + \cdots) \\
&= (1-p)\frac{d}{dp}\left(\frac{p}{1-p}\right) \\
&= (1-p)\frac{1}{(1-p)^2} \\
&= \frac{1}{1-p}.
\end{aligned}
$$

For example, suppose that each B_i has equal probability of being larger or smaller than A; that is, $p = (1-p) = \frac{1}{2}$. This may be a reasonable assumption to make in the absence of any specific knowledge about the distribution of values in the tree. Then, according to the analysis above, we may expect an average distance to cutoff to be no greater than $D(\infty, \frac{1}{2}) = 2$, regardless of how many B_i there are.

3.3.2. Approximate Evaluation of Game Trees

In the preceding section we showed how to evaluate any game that always ends and whose rules assign a value (payoff) to each terminal position. The reason that this does not yield a practical algorithm is that, for all but the most trivial games, the tree to be evaluated is so large that no computer could finish the evaluation in a reasonable time. The reader is invited to check this assertion on any of the games he knows, and using the speed of operation of any computer, existing or foreseeable. Hence in the design of programs for playing nontrivial games, we must give up the goal of optimal play, and strive instead to approximate this by reasonably good play. This is achieved by replacing the exhaustive search of the entire game tree, which is implicit in the preceding discussion, by a *selective search* of a small portion of the game tree. We try to apply common sense and experience with the game at hand to ensure that the part of the tree which is searched is relevant, that is, that it contains the good moves and avoids most of the poor ones. For this reason, this is also known as a *heuristic search* of the game tree. The purpose of this section is to describe how such heuristic searches may be carried out.

One key notion involved in heuristic searches is that of a *static evaluation function*, which assigns to any position P a static value, $s(P)$, which is, one hopes, a good approximation to the theoretical value $v(P)$ defined in the preceding sections. The word static reflects the fact that the value $s(P)$ is obtained by analyzing the position P only, without generating any of its successors. This contrasts with the theoretical value $v(P)$, which is obtained from the values of all the successors of P, which are obtained in turn from the values of their successors, and so on, down to the terminal nodes. The purpose of the static evaluation function is to be able to start the process of tree evaluation from nodes other than terminal ones, because, as, we said earlier, terminal nodes are, in general, too far down in the tree.

How does a static evaluation function assign a value to a position? Here is the one point in the entire scheme of heuristic tree searching where a dependence on the particular game to be played cannot be avoided. For all known games, experience has shown that there are certain features which are closely correlated with a good position, and others with a bad position. Such features may be as simple to compute as the excess number of pieces I have over my opponent (as in checkers), or as difficult to recognize as "a favorable end game" in chess. The point to remember is that without some specific knowledge about the game to be played there is no way to design a static evaluation function which is better than random. However, for all games that have been considered, there have been static evaluation functions which can be computed in a practical way and which give reasonable approximations to the theoretical value in most positions.

When a reliable static evaluation function is available, a game-playing program can be easily designed: from any position P, move to that successor of P which has the highest static value. It is plausible, and experience has confirmed this, that this method leads to poor game-playing programs. Static evaluation functions can recognize features that are associated with good positions most of the time, but hardly any are perfect. For example, probably every static evaluation function that has been used for chess or checkers had a term corresponding to the material advantage (difference in the number of pieces) of one player over the other. It is clear that this term will contribute little which is reliable, and may lead to an erroneous static value, during an exchange of several pieces or during a sacrifice of pieces done with the purpose of capturing enemy pieces later.

Once we recognize that it is usually possible to find reasonable static evaluation functions which may be wrong in some positions, we are naturally led to trying to combine the static values of neighboring positions in such a way that the unreliable values are eliminated. In addition, we want to use the reliable ones to determine a *heuristic value* h(P), which is a better approximation to the theoretical value v(P) than the static value s(P) is. A few wrong static values can usually be made harmless, as long as many of their neighboring positions have reasonably accurate static values, since game trees are evaluated by the minimax process described in the preceding section.

The example in Figure 3.12 illustrates this point. Assume that position P and all its first- and second-level successors have a theoretical value of 0 (this might mean that the game is even). Now assume that we apply a static evaluation function to all second-level successors of P, and that for most of these we obtain the correct static value of 0; but for a few we get the wrong values 9 or −9. In a minimax evaluation of the tree, a min level weeds out the values that are too large, and a max level weeds out the ones that are too small.

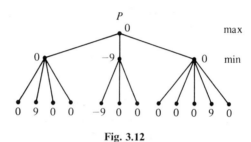

Fig. 3.12

This suggests a procedure for evaluating game trees known as *fixed-depth search*. We present it not as an example of a good search procedure, but in order to review the concepts involved. Given a static evaluation function s, which gives reasonably good approximations s(P) to the theoretical value v(P) for most positions P, compute a heuristic value h(P) by first computing s(P′) for all descendants P′ of P d levels below P, where d is a parameter usually determined by the computation time available to obtain h(P). Now apply the minimax evaluation procedure to all the computed successors P′ of P to obtain h(P).

Many early game-playing programs used fixed-depth search, but nowadays this scheme has largely been abandoned. Fixed-depth search has two disadvantages:

1. Time is wasted by searching foolish moves. In most games and real-life decisions, there are an enormous number of moves to consider. Most of these, however, can be almost immediately discarded (i.e., after a shallow search) as being too unpromising to warrant further effort. By pursuing unpromising branches to the same depth as promising ones, fixed-depth search prevents the latter from being explored more thoroughly, since the total search time is limited.

2. The positions P' whose static value $s(P')$ is computed are chosen without any regard to whether or not the static evaluation function is likely to be reliable on these positions. If P' is, for instance, in the middle of an exchange of pieces in chess, it may well be that the static value of some predecessor of P' would have been more reliable. In such a case, the quality of the search could have been increased while decreasing the search time.

These considerations lead to two important ideas: the first to that of a *selection operator*, and the second to that of a *quiescent position*. By a selection operator we mean a function σ, which, when applied to a position P, selects out of the set of successors of P a subset which it considers to be promising, that is, those worth further investigation. A selection operator is of course specifically tailored to the game being considered, but this need not be explicit; it can be, and frequently is, defined in terms of the static evaluation function. It is convenient, and quite realistic, to think of the static evaluation function as the only place in a heuristic search scheme where game dependence appears explicitly. We shall do so, and hence we only have to consider how the selection operator can be implemented in terms of the static evaluation function. One reasonable way is to assign to each successor P' of P a value $h(P')$, which is obtained by carrying out a fixed-depth search to a shallow depth d beginning at P' [then the special case $d = 0$ uses the static value $s(P')$ for selection], and then to select a fixed number k of positions P' that have the highest values $h(P')$.

The intuitive notion of a *quiescent position* is one in which "nothing wild is happening," which can be judged on the basis of common sense and experience with the game, without having to worry that a neglected detail may render the judgment worthless. The opposite of a quiescent position is one that is full of hidden threats. Our concern with quiescent positions is due to our desire to detect those positions where the static evaluation function gives reliable values. Hence, to be precise, the relevant notion is that of a quiescent position *with respect to* a given static evaluation function. We would like to use these positions to terminate the search, in order to have reliable static values to use as the starting values for the backup procedure.

How can one detect positions whose static value is reliable? This is impossible in general, of course, without a dynamic analysis of the position. It is possible in many cases, however, to detect when the static value of a position is unlikely to be reliable. Assume, for instance, that for some path from the root of a game tree to a node, the static values of the positions keep alternating between being very favorable and very unfavorable. This might be the case for a prolonged exchange of pieces in chess with a static evaluation function that measures primarily the number of pieces each player has. Clearly, no static value along such a path can be assumed to be reliable;

we must look farther down the tree until the static values have settled down. In our example from chess, this may be expected to happen when no pieces have been taken for several moves.

Thus we are led to say that a position is quiescent if its static value "agrees" with the static values of neighboring positions. This vague notion of agreement can be made more precise in various ways: we might say that position P is quiescent if its static value $s(P)$ differs from the heuristic value $h(P)$ obtained from a fixed-depth search (to some shallow depth d) by less than a given tolerance. Such a definition of quiescence in terms of the static evaluation function means that the only explicit game-dependent feature in the search scheme enters through the static evaluation function, as was the case with the selection operator. There are other reasonable ways to define a selection operator and quiescent positions. In order to discuss a variable-depth heuristic search procedure in some generality, we shall from now on only consider a selection operator σ, which selects certain successors $P', P'', \ldots$ of any position P, and a quiescence test q, where $q(P)$ is true if P is quiescent and is false otherwise.

There is one more aspect of heuristic tree search that is worth generalizing, the method by which values are "backed up" from lower nodes to higher nodes in the tree. So far we have assumed tacitly that this is the minimax procedure discussed earlier, which is based on the theory of games; but heuristic search is useful also in trees that are not game trees, and then the minimax backup procedure need not apply. Even in game trees, there are other backup procedures worth considering. As an example, let us mention Slagle's *m and n backup procedure*, which is based on the following idea. If the values to be backed up were exact, that is, the theoretical values of positions, then the minimax backup procedure would be best; however, all values we actually operate upon are inexact and occasionally subject to gross errors. As a consequence, a cautious player may prefer the position P' in Figure 3.13 to the position P'', even though the latter has a higher value if the

Fig. 3.13

minimax backup procedure is used (assuming that P' and P'' are at a max level). The reason for this is that in P' there are several reasonable possibilities, corresponding to the values 8, 6, and 5, while in P'' there is just one. If this value 9 should turn out to be wrong, he might find himself in an undesirable situation when he gets to P''. Remember that a player is concerned with the values of P' and P'' before he actually gets there, and so looking at the positions P' and P'' from several moves earlier, when one can still postpone the actual decision as to which move to make from P', P' may look like a much more desirable goal than P'', since it is safer. The m and n backup procedure takes this into account by assigning to a position at the max level the average of the m largest values of its successors, and to a position at the min level the average of the n smallest values of its successors. The special case of 1 and 1 search is the minimax procedure.

Let us denote a backup procedure by β. It assigns to a position P a value $\beta(x', x'', \ldots)$, which depends on the values $x', x'', \ldots$ obtained for the successors $P', P'', \ldots$ of P, and in general also depends on the level of P. With these concepts introduced, we can now concisely describe a *variable-depth search* procedure. Given a static evaluation function s, a selection operator σ, a quiescence test q, a backup operator β, and a cutoff depth d, compute a heuristic value $h_d(P)$ as follows:

$$h_d(P) = s(P) \qquad \text{if } d = 0 \text{ or if } q(P) \text{ is true,}$$
$$h_d(P) = \beta(h_{d-1}(P'), h_{d-1}(P''), \ldots) \qquad \text{where } P', P'', \ldots$$
$$\text{are all the successors of } P \text{ in } \sigma(P).$$

The introduction of a cutoff depth d is necessary to stop the search, even if the quiescence test q should fail to stop it. The next section shows how such a search technique can be applied to a specific game.

3.3.3. An Example: Fast Wins in Shannon Switching Games

Let us illustrate these ideas on game tree evaluation on an actual game. A slight modification of the Shannon switching game discussed in Section 3.1.2 provides an example that is readily explained and has the additional advantage that no mathematical solution is known other than an exhaustive search of all possible moves, which would mean constructing the entire game tree. We are now interested not only in whether and how a player can win, but how he can win fast, and thus there is the dual problem of how the player who must lose can delay his defeat as long as possible. We call this the problem of *minimum-time wins* and *maximum-time losses* in Shannon switching games.

It is interesting to note that the theory of Shannon switching games developed in Section 3.1.2 does not apply here. Moreover, a Short player who plays according to a two-tree strategy will tend to win rather slowly since this strategy is designed to keep all the nodes connected that are spanned by the two trees; it lacks a sense of direction for attempting to connect only the two distinguished nodes.

An example of this is shown for a Short game in Figure 3.14(a), where the graph has been decomposed into a tree drawn with straight lines and

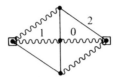

(a) The initial configuration

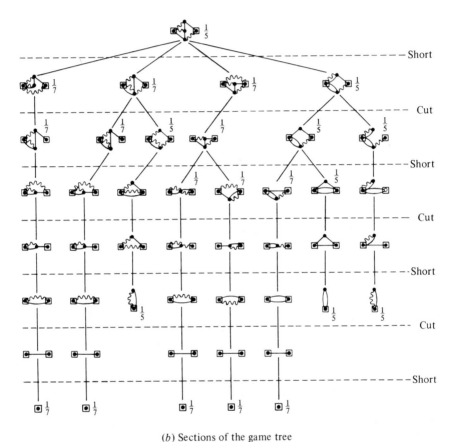

(b) Sections of the game tree

Fig. 3.14 A fast win in a Shannon switching game.

another one drawn with wiggly lines. If Cut deletes edge 0, the strategy based on these two trees will make Short claim edge 2, and will thereafter guide him to a win that requires that all eight edges of the graph be claimed by one or the other opponent. However, an analysis of the game tree which arises after Cut deletes edge 0 shows that if Short claims edge 1 as his first move he can win faster. Sections of this game tree are shown in Figure 3.14(b).

The values that have been assigned to the leaves (terminal nodes) of the tree need an explanation. Zero means that Short loses; a positive number means that Short wins. Ordinarily, values assigned to terminal positions depend only on these positions, and not on the history of how they arose. However, if we want to give a higher payoff for a fast win than for a slow win, the payoff must depend on the number of moves the game lasted. The simplest choice is to make the payoff of a terminal node inversely proportional to the level of that node, which is defined as follows: the level of the root is zero, and the level of any other node is 1 plus the level of its immediate predecessor. Thus the level of a node in a game tree is equal to the total number of moves and countermoves that were made in getting from the starting position to the position represented by this node.

Let us now consider the crucial problem of finding a reasonable static evaluation function for this game. This always results in a compromise between

1. A sophisticated static evaluation function, which presumably gives a good approximation to the theoretical value of a position, but which is rather time consuming, and hence limits the depth to which the tree can be searched, and

2. A simple static evaluation function, which is not very reliable but fast, and may make up for its poor judgment of a position by the fact that it allows a deeper search in a fixed length of time.

As a rule of thumb, it is worthwhile to use a sophisticated static evaluation function in a game in which one has collected a lot of experience, so that one can be sure that the features which enter into the static evaluation function are relevant to the game. A static evaluation function that checks many features of a position is likely to be a waste of time if one is not sure which features should be examined. Since we do not have a good understanding of the problem of fast wins on Shannon switching games, we propose the following simple static evaluation function. If Short wants a fast win on a Shannon switching game, it is natural that he be interested in the minimal number of edges which he must claim to connect the two distinguished nodes, assuming no interference by Cut. Call this distance between the two distinguished nodes δ. It is also plausible that, other things being equal, the smaller the δ, the more favorable a position is for Short. This suggests a static evaluation function $s(P)$ that is inversely proportional to the distance δ between the two distinguished nodes in position P, $s(P) = 1/\delta$. There is a conceptual difficulty, however, since we have defined the theoretical value $v(P)$ of a terminal position P to be inversely proportional to the level l of P. Hence, if our static evaluation function is to be a good approximation to v, it should

depend on the level l of the position P in the game tree also. A static evaluation function of the form

$$s'(P) = \frac{1}{l + 2\delta}$$

achieves this purpose. Notice that it agrees with the theoretical value $v(P)$ for
terminal positions P, where $\delta = 0$ if Short won, and $\delta = \infty$ if Cut won. In a
game where Short lessens the distance between the two distinguished nodes
by 1 in each exchange of move and countermove, $s'(P)$ agrees with the theoretical value throughout.

In practice, either s or s' leads to the same moves. We have made
numerous experiments with these static evaluation functions and found that
they perform quite well on graphs chosen at random if the cutoff depth of the
tree search is not too shallow. However, it is clear that there are quite a number of graphs on which the distance δ between the two distinguished nodes is
unrelated to the number of moves it takes for Short to win. The problem of
finding fast wins on Shannon switching games is an open one, and we pose it
primarily to provide an example on which the interested reader can experiment with the techniques of heuristic search.

3.4. REMARKS AND REFERENCES

Shannon's early paper on a chess-playing program is

SHANNON, C. E. "Programming a Digital Computer for Playing
Chess," *Phil. Mag.*, *41* (1950), 356–357.

Mack Hack VI, the program that achieved a significant step forward in computer chess-playing ability, is described in

GREENBLATT, R., D. EASTLAKE, and S. CROCKER. "The Greenblatt
Chess Program," *AFIPS Conference Proceedings*, Vol. 31
(FJCC 1967), pp. 801–810, Thompson Books, Washington,
D.C., 1967.

Samuel's extremely successful checker-playing program is described in
two papers written eight years apart. These give a good insight into the
progress in heuristic programming that had been achieved in the years between
these two publications. They are

SAMUEL, A. "Some Studies in Machine Learning Using the Game
of Checkers," *IBM J. Res. Develop.*, *3* (1959), 210–229.

SAMUEL, A. "Some Studies in Machine Learning Using the Game of
Checkers, II—Recent Progress," *IBM J. Res. Develop.*, *11*
(1967), 601–617.

The main theorem on Shannon switching games is due to Lehman in

LEHMAN, A. "A Solution to the Shannon Switching Game," *SIAM J. Appl. Math.*, *12* (1964), 687–725.

For an example of how a computer can be programmed to play Shannon switching games, see

CHASE, S. M. "An Implemented Graph Algorithm for Winning Shannon Switching Games," *Comm. ACM*, *15* (1972), 253–256.

The game of calling the largest number was invented by J. H. Fox, Jr., and L. G. Marnie; the analysis given in Section 3.1.3 is due to L. Moser and J. R. Pounder. Both the game and its analysis were first published in the Mathematical Games section of *Scientific American* in February and March 1960. The analysis we presented shows only that of all strategies which examine an initial fraction of the numbers the fraction $1/e \approx 0.37$ is best. E. B. Dynkin has shown that this strategy is best over *all* possible strategies. This is a deep mathematical result; for the details see

DYNKIN, E. B. "The Optimum Choice of the Instant for Stopping a Markov Process," *Dokl. Akad. Nauk SSSR*, *150* (1963), 238–240 (Russian). An English translation appears in *Soviet Math.*, *4* (1963), 627–629.

It turns out that this game suggests a useful method of dynamic storage allocation, a difficult real-life computer problem. See

CAMPBELL, J. A. "A Note on an Optimal-Fit Method for Dynamic Allocation of Storage," *Computer J.*, *14* (1971), 7–9.

The game of Hex was introduced in the 1940s independently by Piet Hein of Denmark and John Nash of the United States. For more details on Hex, see Chapter 8 of

GARDNER, M. *Mathematical Puzzles and Diversions*, Simon and Schuster, New York, 1959.

The birth of the theory of games is usually attributed to the paper

VON NEUMANN, J. "Zur Theorie der Wirtschaftsspiele," *Mathematische Annalen 100* (1928), 295–320.

The first book on the subject,

VON NEUMANN, J., and O. MORGENSTERN. *Theory of Games and Economic Behavior*, Princeton University Press, Princeton, N.J., 1944,

stressed the application of game theory to economic problems. A comprehensive survey of game theory and decision-making models can be found in

LUCE, R. D., and H. RAIFFA. *Games and Decisions*, Wiley, New York, 1957.

The fictitious-play algorithm for finding the value of a game is due to

G. W. Brown, and its convergence is proved in

> ROBINSON, J. "An Iterative Method of Solving a Game," *Ann. Math.*, *54* (1951), 296–301.

A detailed discussion of tree searching and many references to technical papers on the subject are contained in

> SLAGLE, J. R. *Artificial Intelligence—The Heuristic Programming Approach*, McGraw-Hill, New York, 1971.

An assessment of the progress achieved in heuristic programming over the past two decades can be found in

> SIMON, H. A. "The Theory of Problem Solving," *Information Processing 71*, C. V. Freiman (ed.), Vol. 1, 261–277, North-Holland, Amsterdam (1972).

It is worth noting that the problem of tree searching has many applications other than in game playing. An important example is discussed in

> SLAGLE, J. R. and R. C. T. LEE. "Application of Game Tree Searching Techniques of Sequential Pattern Recognition," *Com. ACM 14* (1971), 103–110.

3.5. EXERCISES

1. Analyze the version of Nim in which the player who moves last loses instead of wins.

2. Consider a Shannon switching game on a planar graph, that is, a graph which can be drawn on a sheet of paper without crossing edges. Denote the two distinguished nodes by squares. Assume that when a single nonplayable edge (dotted in the figure) is added which joins the two distinguished nodes, without destroying planarity, then the set of edges can be partitioned into two spanning trees (one drawn with wiggly edges in the figure, the other with straight edges, including the dotted edge).

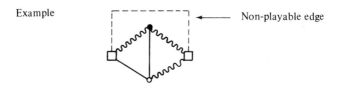

Example   Non-playable edge

The dual graph of a planar graph is obtained by drawing a node in each of the regions of the original graph, including the outside region, and by joining two nodes with an edge if the corresponding regions are adjacent, that is, share an edge of the original graph as a common boundary. The two distinguished nodes of the dual graph are those whose regions share the nonplayable edge as boundary. The dual graph for our example is shown with double lines.

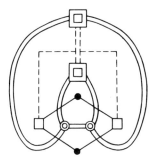

Associate with each edge of the original graph that edge of the dual graph which it intersects. In particular, this associates with the nonplayable edge of the original graph a nonplayable edge of the dual graph. Convince yourself that the following statements are true.

(a) Duality is a symmetric relationship; that is, any planar graph is a dual of its dual.

(b) It is not true in general that a graph and its dual are isomorphic, as our example might have suggested.

(c) Any Shannon switching game of the type discussed generates, by means of the one-to-one correspondence between its edges and edges on the dual graph, a game on the dual graph in such a way that if Short wins in the original game, Cut wins in the dual game, and vice versa.

(d) A graph gives a neutral game if and only if it has two edge-disjoint spanning trees only after the addition of an edge joining its two distinguished nodes.

3. Analyze the following versions of the game "calling the largest number." The assumptions are the same as those of Section 3.1.3, unless stated differently.

Version 1: Calling the number closest to a given number z.
   Given a number z before the game starts, the oracle should identify that number in the shuffled sequence which is closest to z.

Version 2: Calling the kth largest number.
   Given an integer k, $1 \leq k \leq n$, before the game starts, the oracle should identify the kth largest number in the shuffled sequence. Of special interest is the case of the median, where $k = [n/2]$.

What are the best strategies for the oracle that you can devise?

4. Find the value and the minimax strategies for the two-person zero-sum games with the following payoff matrices.

$$\begin{pmatrix} 1 & 2 & 9 \\ 4 & 3 & 4 \end{pmatrix}, \quad \begin{pmatrix} 1 & 2 & 3 \\ 3 & 2 & 1 \end{pmatrix}.$$

5. Write a program that plays two-person zero-sum games against itself according to the method of fictitious play discussed in Section 3.2.2. Determine experimentally the rate of convergence with which this method finds the value of a game.

6. For some two-person zero-sum games with which you are familiar, estimate
   (a)   The number of strategies each player has.
   (b)   The number of nodes in the game tree.

7. For the Shannon switching game illustrated in Exercise 2, compute exactly
   (a)   The number of nodes in the game tree.
   (b)   The number of strategies each player has.

8. Describe how the number of strategies a player has can be computed from the game tree.

9. Find two orderings of the nodes of the tree of Figure 3.8 that maximize and minimize the efficiency of the alpha–beta pruning procedure.

10. For some game with which you are familiar (e.g., fast wins on Shannon switching games), experiment with various choices for a static evaluation function, a selection operator, and a quiescence test, as discussed in Section 3.3.2.

11. Write a program that generates a game tree, assigns arbitrary values to the terminal nodes, and evaluates this tree with the minimax backup procedure. Suppose that some terminal values are modified (random errors are introduced). Have your program reevaluate the tree using the minimax backup procedure as well as the m and n backup procedure, for various values of m and n. Compare the results obtained by the various backup procedures in the presence of errors with the theoretical value.

12. Using whatever techniques from this chapter you find appropriate, develop a computer program that plays a perfect game of tic-tac-toe.

4 RANDOM PROCESSES
ON DETERMINISTIC COMPUTERS

Most computers are deterministic devices whose results are, assuming no errors, predictable and reproducible: the same program with the same data will produce the same results. If there is anything random about computers, it is that they sometimes operate in an unpredictable environment, so that one cannot tell what data they will receive. On the other hand, most real-life phenomena have unpredictable components; for example, the traffic at an intersection at a given time of day may follow certain statistical patterns, but it is subject to random fluctuations. Such fluctuations prevent us from making an *exact* mathematical analysis; all we can do is give a probabilistic analysis. One might suspect that a deterministic device cannot, by its very nature, be used to simulate random phenomena. However, this is commonly done, and this chapter describes some of the techniques used to achieve such a simulation.

First, we shall consider some of the aspects of randomness that we want to simulate, and then we shall propose a method to effect such a simulation. Having created pseudorandom behavior, we shall show how to measure how well we have approximated random behavior. Finally, we shall examine various ways of using this behavior. Such usage is not restricted to studying statistical phenomena; we can also use random behavior to find areas and volumes, or to compute the value of π.

4.1. THE MEANING OF RANDOM

In everyday language we usually use the word "random" to mean unpredictable or unknown. Thus if someone watches the arrival of buses at a station without knowing the schedule, he might say that the buses arrive at random times. After he has detected the daily pattern, he would say only that unexpected deviations from the schedule occur at random. This aspect of randomness, its unpredictability, is not commonly used to give a precise mathematical definition of "random." To state such a definition with the needed rigor, we would have to state by whom, by what means, and under what conditions an event is unpredictable. Such a definition would need revision for each scientific advance that allowed the prediction of a previously unpredictable event. For example, to primitive man the weather on a given day must have appeared random and unpredictable, whereas using modern meteorological techniques, weather can be forecast with considerable accuracy. Hence we shall look for an alternative characterization of randomness.

We shall develop such a characterization by means of an example. Suppose we were told that a sequence of 0's and 1's was the record of a sequence of tosses of an unbiased coin, the kth digit indicating the outcome of the kth toss (0 for heads, 1 for tails). How would we decide whether or not this sequence actually did occur as the result of tossing a fair coin? Even though we could never be certain whether the sequence arose in that manner, we might make a subjective judgment. We would probably reject the sequence

$$0100000100000000000001100$$

since it contains a disproportionate number of 0's; certainly one thing we would expect is approximately the same number of 0's and 1's. But what about

$$101010101010101010101010$$

or

$$111111111111000000000000?$$

They certainly satisfy that criterion, and yet they seem much too regular to have arisen in a random manner.

How can we express the departure from irregularity that the sequences exhibit? Consider all pairs of consecutive digits, and see how frequently the four possible patterns occur in each sequence:

Pattern	Number of Occurrences in First Sequence	Number of Occurrences in Second Sequence
00	0	11
01	11	0
10	12	1
11	0	11

Thus our second criterion should be that each combination of two consecutive digits occurs with approximately equal frequency. By considering

$$001100110011001100110011$$

we find that though our two criteria are met, the sequence appears far from random. From these examples it is not surprising that for a sequence to be considered random we must have, for each $k > 0$, all the 2^k different patterns of length k occurring with approximately equal frequency. We have used the term "approximately" in a vague sense, but it should be clear that a measure could be defined which determines how well a given sequence satisfies the stated criteria.

Our criteria may be fine for infinite sequences, but notice that no finite sequence can satisfy them. If a sequence is k digits long, then only one pattern of length k will occur, and the other $2^k - 1$ will not. Thus it is important that we have a rigorous definition of how well the criteria are satisfied for finite sequences; there is no "perfect" finite random sequence using these criteria.

Are there "perfect" infinite random sequences? The answer is yes, and, moreover, some very regular looking sequences satisfy all the criteria. Consider the sequence

$$1101110010111011110001001101010111\ldots,$$

which is formed by placing the binary representations of the integers next to each other, in order: 1, 10, 11, 100, 101, 110, 111, 1000, and so on. It can be shown that for all values of k and all blocks $B_k = b_1 b_2 \ldots b_k$ of length k

$$\lim_{n \to \infty} \frac{N(n, B_k)}{n - k + 1} = \lim_{n \to \infty} \frac{N(n, B_k)}{n} = \frac{1}{2^k},$$

where $N(n, B_k)$ is the number of occurrences of the block B_k in the first n digits of this sequence. Similarly, the sequence

$$1011101111011\ldots,$$

formed by placing the binary representations of the prime integers next to each other in order, has exactly the same property! These are hardly "random sequences" in the usual sense, and so it seems that unpredictability may no longer be a necessary requirement for randomness.

We shall not now complete our definition of randomness by attempting to define a measure of how well a finite sequence satisfies the stated criteria; this topic will be discussed in Section 4.1.2. Instead, let us summarize our discussion by saying that any workable definition of randomness has to be based on a number of tests which can be administered objectively and which can be evaluated as to how well each test is satisfied in any particular case. Furthermore, there must be sequences (like the two previous examples) that can be easily computed and also satisfy the chosen tests. Such a sequence of digits can be generated by a deterministic computer, and yet the sequence can be used as a source of randomness in computer simulations of real-life phenomena. In addition, since these sequences are reproducible, we can repeat exactly the same "random" calculation in order to check for certain intermittent errors that might have occurred in the computer.

Computations that require a source of randomness are a significant part of computer applications in the physical sciences and in operations research. Deterministic processes are usually used in such cases as a source of randomness; there is no contradiction as long as we keep in mind that randomness can be defined mathematically by the frequencies of occurrence of patterns.

4.1.1. Random-Number Generators

A *random-number generator* (sometimes called a *pseudo-random-number generator*) is an algorithm that generates (defines) a sequence

$$r_1, r_2, r_3, \ldots, r_k, \ldots$$

of numbers which "appears to have resulted" from successive independent observations of the values of a random variable. The phrase "appears to have resulted" must be defined in each particular case. For example, suppose that we want to simulate the random variable R, which assumes integer values uniformly distributed in the interval $[0, 8191]$; in other words every integer in this interval has equal probability, $1/8192$, of occurring as a value of R. Any sequence of integers between 0 and 8191 is a possible outcome of successive observations of R, and, moreover, any sequence is as likely to occur as any other sequence of the same length. Thus there is no such thing as a "most likely sequence" that we could use for a random-number generator. However, not all patterns in sequences are equally likely to occur; for instance, the probability of a sequence in which integers less than 10 occur ten times as often as other integers is much smaller than the probability of a sequence in which every integer occurs equally often. Why? The reason is simple: there are more sequences of the latter type than of the former type, and so sequences of the former type are relatively improbable. Thus what we mean by saying that a sequence appears to have arisen from successive observations of a random variable is that it does not contain improbable patterns.

Which patterns are to be considered improbable? This part of the definition will differ for each specific problem. For our example, we shall require that the frequency with which random numbers fall into subintervals of equal length be approximately equal. More specifically, divide the interval $[0, 8191]$

into 16 subintervals, each consisting of 512 consecutive integers, that is, [0, 511], [512, 1023], . . . , [7680, 8191]. For a given subinterval let $O(k)$ be the number of elements of the sequence $r_1, r_2, \ldots, r_k$ which are contained in that subinterval. We want to bound the difference

$$D(k) = O(k) - \frac{k}{16},$$

which is the discrepancy between the actual number of r_i's in the subinterval and the expected number. It is unreasonable to ask that $D(k)$ be bounded by a constant, independent of k, since we know from experience that in tossing a coin k times the number of heads can differ from the number of tails by an arbitrarily large amount, if k is large enough. Using probability theory, it can be shown that such fluctuations, measured by the standard deviation, are expected to be proportional to the square root of k; for a derivation of this fact see Exercise 2. So, for each subinterval we shall require that

$$|D(k)| < c\sqrt{k},$$

for some constant c; but we shall not assign a specific value to c, since the asymptotic behavior given by the factor $\sqrt{k}$ is really the relevant aspect. In practice, most random-number generators would be required to satisfy more conditions than this one; the generator described later does indeed satisfy more stringent requirements.

Since a random-number generator is usually called many times, it must be fast; thus the process of generating a new number must be simple and efficient. We could, of course, compute a large number of terms in the sequence and then store them in some permanent form. This was, in fact, the way it was done in the days of mechanical card tabulating machines. A sequence of random numbers was computed and punched onto cards; then, every time another random number was required, a new card would be read. Such a method was acceptable when the input/output speeds were comparable with the speeds of arithmetic operations; but today's computers can add, subtract, multiply, and divide a thousand times faster than they can read from external sources or backup storage devices, so it would be very slow to precompute the numbers and read them as needed. It would also be wasteful in terms of space, since the random sequences needed are usually quite long. It is more efficient to recompute the series each time it is needed.

One random-number generator that satisfies our requirements of frequency, speed, and simplicity is defined by

$$r_1 = 1,$$
$$r_{k+1} = 125 r_k \ (\text{modulo } 8192), \qquad k \geq 1.$$

Obtaining r_{k+1} from r_k involves a multiplication and finding the remainder. Since $8192 = 2^{13}$, finding the remainder is done by simple shift operations on a binary computer; a similar generator can be devised for a computer that represents numbers in decimal notation.

Let us examine some of the properties of this sequence. Since every number generated is an integer in [0, 8191], some integer must be repeated. Assume that $r_i = r_{i+l}$, for some $l > 0$. From the definition of the sequence, it is clear that for $k \geq i$, $r_k = r_{k+l}$; that is, the sequence will repeat with period l. A periodic sequence certainly does not conform to our intuitive notion of randomness, but the periodicity will not affect anything unless the sequence is used for at least the length of the period. Notice that if we choose 2 as a multiplier instead of 125 then the period would be 1, since $r_{14} = 0$ (why isn't the period 14?). Using 125, the period is 2048, so that exactly one quarter of the integers in [0, 8191] occur in the sequence.

Every r_k has the form $4n + 1$, and every number of this form in [0, 8191] will occur in the sequence. This is, of course, an unlikely pattern, which would render the sequence unacceptable for many applications. However, we can use this generator to obtain uniformly distributed real numbers in the interval (0, 1) by scaling, that is, dividing each r_k by 8192; from this point of view the "parity" of r_k is irrelevant. Again, this points out the fact that there is no such thing as a "universally good random sequence"; the "goodness" of a random sequence depends on the intended application.

To discover how these r_k are distributed in the various intervals, a computer program was written to compute these numbers and examine their occupancy in the 16 subintervals. The results of that program are given in Table 4.1. Notice that the last column is roughly constant, supporting the assumption that the sequence obeys a *square-root law*.

4.1.2. How Random Is Random?

How can we quantitatively measure how close we have come to obtaining random behavior? Given a random-number generator we can, by theoretical analysis, study the behavior of the generator, or we can apply various empirical tests and evaluate the results. This section describes several well-known empirical tests and how their results are evaluated; a detailed theoretical analysis of even the simple generator given in the previous section is beyond the scope of this book.

Actually, we have already introduced one empirical test, the *frequency test* described in the previous section. In examining Table 4.1 for the values $k \geq 100$, we see that the values of r_i seem to be nicely distributed over the 16 subintervals; but how do we *quantitatively* measure such a distribution? The method usually used is called the *chi-square test*. To explain this test, consider the following simple example. Suppose that we have a *biased* coin which lands

Table 4.1 DISTRIBUTION STATISTICS FOR A SIMPLE RANDOM-NUMBER GENERATOR

k	r_k	Occupancy of Intervals																$\min \frac{D(k)}{k}$	$\max \frac{D(k)}{k}$	$\max \frac{|D(k)|}{\sqrt{k}}$
1	1	1	0	0	0	0	0	0	0	0	0	0	0	0	0	0	0	−0.06	0.94	0.94
2	125	2	0	0	0	0	0	0	0	0	0	0	0	0	0	0	0	−0.13	1.88	1.33
3	7433	2	0	0	0	0	0	0	0	0	0	0	0	0	0	1	0	−0.19	1.18	1.05
4	3429	2	0	0	0	0	0	0	0	1	0	0	0	1	0	1	0	−0.25	1.75	0.88
5	2641	2	0	0	0	0	1	0	1	0	0	0	0	1	0	1	0	−0.31	1.69	0.75
6	2445	2	0	0	1	1	1	0	1	0	0	0	0	1	0	1	0	−0.38	1.63	0.66
7	2521	2	0	0	1	1	1	1	1	0	0	0	0	1	0	1	0	−0.44	1.56	0.59
8	3829	2	0	0	1	2	1	1	2	0	0	0	0	1	0	1	0	−0.50	1.50	0.53
9	3489	2	0	0	1	2	2	1	2	0	0	1	0	1	0	1	0	−0.56	1.44	0.48
10	1949	2	0	0	1	2	2	2	2	0	1	1	0	1	0	1	1	−0.63	1.38	0.43
20	421	3	1	1	2	2	3	0	4	2	1	1	0	0	0	1	1	−1.25	1.75	0.39
30	2285	3	1	1	2	4	4	1	4	3	2	1	0	1	0	1	2	−1.88	2.13	0.39
40	3957	4	1	1	2	2	5	3	4	3	3	2	1	1	1	1	3	−2.50	2.50	0.40
50	5949	5	1	1	3	6	2	3	7	3	3	2	0	3	1	1	3	−3.13	3.88	0.55
60	4677	5	1	3	3	6	3	3	7	4	7	3	1	3	1	2	3	−2.75	5.25	0.68
70	653	5	3	3	3	7	9	3	7	7	3	4	1	2	2	3	4	−2.38	6.63	0.79
80	6677	5	4	3	4	7	11	3	7	7	5	5	2	3	3	4	6	−3.00	6.00	0.67
90	6877	5	4	4	3	7	11	4	7	7	5	5	3	6	6	6	6	−2.63	5.38	0.57
100	5861	6	4	4	7	10	12	4	7	7	7	7	7	8	6	8	6	−2.25	5.75	0.57
200	4597	12	9	8	14	15	15	11	19	26	16	17	10	15	18	11	12	−5.50	5.50	0.39
300	6661	18	14	15	23	19	19	14	24	26	19	26	17	23	25	17	14	−4.75	7.25	0.42
400	7957	24	19	24	30	19	26	18	30	33	21	26	24	34	34	24	28	−7.00	9.00	0.45
500	4389	31	23	28	35	23	33	24	34	41	26	32	32	43	37	32	33	−8.25	11.75	0.53
600	53	39	28	36	42	30	41	28	44	43	34	36	36	46	41	36	47	−9.50	9.50	0.39
700	7237	47	39	41	48	35	44	34	48	41	43	52	42	50	46	42	48	−9.75	8.25	0.37
800	5461	50	47	47	49	41	47	52	55	51	55	50	46	57	50	50	50	−11.00	10.00	0.39
900	7013	57	56	53	62	48	56	56	60	56	60	63	55	67	57	55	54	−10.25	10.75	0.36
1000	7796	68	64	62	69	56	69	55	58	64	60	64	60	69	56	65	58	−7.50	6.50	0.24

147

heads up with probability $p_h = \frac{1}{3}$ and lands tails up with probability $p_t = \frac{2}{3}$. If we toss this coin 100 times, obtaining 36 heads and 64 tails, how closely does this match the expected outcome of $33\frac{1}{3}$ heads and $66\frac{2}{3}$ tails? We could compute the sum of the squares of the differences between the expected results and the observed results. Letting

$$o_t = \text{observed number of tails,}$$

$$o_h = \text{observed number of heads,}$$

and

$$e_t = \text{expected number of tails,}$$

$$e_h = \text{expected number of heads,}$$

we would be computing

$$(e_t - o_t)^2 + (e_h - o_h)^2.$$

Since tails are supposed to occur twice as often as heads, $(e_t - o_t)^2$ will, for a large number of trials, be much larger than $(e_h - o_h)^2$ due to the square-root law mentioned earlier. Thus we want to weigh these two factors differently, and it turns out to be convenient to weight them inversely according to the expected outcomes. So, we compute

$$S = \frac{(e_h - o_h)^2}{e_h} + \frac{(e_t - o_t)^2}{e_t}$$

as a measure of how far our observed values are from the expected values. This value is called the *chi-square statistic*. In our specific case we get $S = 0.32$. Is this an improbable value for S to assume? The answer to this question lies in the fact that when the number of coin tosses per trial goes to infinity, there is a fixed relationship between the value of S and the percentage of trials whose chi-square statistic is expected to be greater than S. So, having calculated this value, we can look up in a table the percentage of trials expected to be even farther from the expected distribution.

In general, suppose that the members of a sequence $s_1, s_2, s_3, \ldots$ fall into k disjoint classes $C_1, C_2, \ldots, C_k$, and $p_i = \Pr(s \in C_i)$, so that the expected number of members of C_i from $s_1, \ldots, s_m$ is $e_i = mp_i$. Suppose that in a given trial we have o_i members in C_i, $\sum_{i=1}^{k} o_i = m$. We want to measure how far this deviates from the expected distribution.

The chi-square statistic is defined by

$$S = \sum_{i=1}^{k} \frac{(e_i - o_i)^2}{e_i}.$$

The relationship between this value and the percentage of trials that are expected to be even farther from the expected outcome is indicated in Table 4.2. This table is used by looking at the line in the table with $v = k - 1$. The value of v is called the number of *degrees of freedom* and is 1 less than the number of classes; it would be too difficult to justify this rigorously, but intuitively we can consider $o_1, o_2, \ldots, o_{k-1}$ to be "independent" of each other, while $o_k = m - \sum_{i=1}^{k-1} o_i$ is not, since $\sum_{i=1}^{k} o_i = m$ is fixed. If the entry for $v = k - 1$ in column p is x, this means that $\Pr(S > x) = p$. In our coin-tossing experiment we have $v = 1$ and $S = 0.32$; looking at the appropriate line of the table we see that $\Pr(S > 0.1015) = 0.75$ and $\Pr(S > 0.4549) = 0.50$, telling us that somewhere between 50 and 75 per cent of the time S will be larger than 0.32. Thus we seem to have had a somewhat random outcome.

Now let us apply the chi-square test to the frequency distributions given in Table 4.1 for the simple random-number generator. For $1000 \geq k \geq 100$ the chi-square values can be easily calculated, since all the expected values are equal. In particular, there are 16 classes (the subintervals), and so the $e_i = k/16$ and the o_i are given in the table. We have

k	Chi-Square Statistic
100	13.2800
200	12.9600
300	13.6000
400	12.3200
500	14.1120
600	13.0666
700	9.2114
800	8.5200
900	7.5911
1000	5.7920

Referring to Table 4.2, the line for $v = 15$, we see that for $100 \leq k \leq 600$ the value of the chi-square statistic will be bigger between 50 and 75 per cent of the time. For $k = 700$, it will be bigger between 75 and 90 per cent of the time, while for $800 \leq k \leq 1000$ it will be bigger over 90 per cent of the time. Thus for the larger values of k the random-number generator produced results that are, in a sense, too well distributed to be random. An explanation of this phenomenon is left to Exercise 4.

Now that we have a way of quantitatively measuring how close a distribution comes to matching some theoretical distribution, we can use this method not only with the frequency test, but with many other tests as well. The number of different tests available is probably as large as the number of different random-number generators: when a new generator is invented that does not perform well on existing tests, the temptation is strong to develop a new test on which the generator scores satisfactorily. We shall discuss one well-known test in the next paragraph and mention another in Exercise 5.

Table 4.2 VALUES OF THE CHI-SQUARE STATISTIC

v/p	99%	90%	75%	50%	25%	1%
1	0.0002	0.0158	0.1015	0.4549	1.3233	2.7055
2	0.0201	0.2107	0.5754	1.3863	2.7726	4.6052
3	0.1148	0.5844	1.2125	2.3660	4.1084	6.2514
4	0.2971	1.0636	1.9226	3.3667	5.3853	7.7794
5	0.5543	1.6103	2.6746	4.3515	6.6257	9.2364
6	0.8721	2.2041	3.4546	5.3481	7.8408	10.6446
7	1.2390	2.8331	4.2549	6.3458	9.0372	12.0170
8	1.6465	3.4895	5.0706	7.3441	10.2188	13.3616
9	2.0879	4.1682	5.8988	8.3428	11.3887	14.6837
10	2.5582	4.8652	6.7372	9.3418	12.5489	15.9871
11	3.0535	5.5778	7.5841	10.3410	13.7007	17.2750
12	3.5706	6.3038	8.4384	11.3403	14.8454	18.5494
13	4.1069	7.0415	9.2991	12.3398	15.9839	19.8119
14	4.6604	7.7895	10.1653	13.3393	17.1170	21.0642
15	5.2294	8.5468	11.0365	14.3389	18.2451	22.3072
16	5.8122	9.3122	11.9122	15.3385	19.3688	23.5418
17	6.4078	10.0852	12.7919	16.3381	20.4887	24.7690
18	7.0149	10.8649	13.6753	17.3379	21.6049	25.9894
19	7.6327	11.6509	14.5620	18.3376	22.7178	27.2036
20	8.2604	12.4426	15.4518	19.3374	23.8277	28.4120

If we were given five independent random decimal digits, we could classify them into poker hands: five of a kind, four of a kind, full house, three of a kind, two pairs, one pair, or a bust. Using the multinomial distribution formula, it is fairly easy to compute the probabilities associated with each of these hands:

$$\text{Pr(five of a kind)} = \text{Pr}(aaaaa) = \binom{10}{1}\binom{5}{5}\left(\frac{1}{10}\right)^5 = 0.0001$$

$$\text{Pr(four of a kind)} = \text{Pr}(aaaab) = \binom{10}{1}\binom{9}{1}\binom{5}{4}\left(\frac{1}{10}\right)^5 = 0.0045$$

$$\text{Pr(full house)} = \text{Pr}(aaabb) = \binom{10}{1}\binom{9}{1}\binom{5}{3}\left(\frac{1}{10}\right)^5 = 0.0090$$

$$\text{Pr(three of a kind)} = \text{Pr}(aaabc) = \binom{10}{1}\binom{9}{2}\frac{5!}{3!1!1!}\left(\frac{1}{10}\right)^5 = 0.0720$$

$$\text{Pr(two pairs)} = \text{Pr}(aabbc) = \binom{10}{1}\binom{9}{2}\frac{5!}{2!2!1!}\left(\frac{1}{10}\right)^5 = 0.1080$$

$$\text{Pr(one pair)} = \text{Pr}(aabcd) = \binom{10}{1}\binom{9}{3}\frac{5!}{2!1!1!1!}\left(\frac{1}{10}\right)^5 = 0.5040$$

$$\text{Pr(bust)} = \text{Pr}(abcde) = \binom{10}{5}\frac{5!}{1!1!1!1!1!}\left(\frac{1}{10}\right)^5 = 0.3024$$

Total = 1.0000

From the above probabilities we can calculate the expected distribution of the seven poker hands. By generating $5n$ random decimal digits we can easily form n random poker hands and then calculate the observed frequencies of the various types of poker hands; we then compare this observed distribution with the expected distribution by means of the chi-square test. This is done for the random-number generator of Section 4.1.1 in Exercise 6.

4.1.3. Transformation of Random Numbers

Like most random-number generators, the one in Section 4.1.1 produces a sequence of uniformly distributed integers modulo n, for some n. By dividing these numbers by $n - 1$, we transform such a sequence into a sequence of real numbers uniformly distributed in the unit interval [0, 1]; from this distribution we can obtain practically any distribution desired. In particular, this section will describe how we can use such a sequence to approximate a normal or Gaussian distribution with any mean and standard deviation. This, the most prominent of all probability distributions, is also known as the bell-shaped distribution. Obtaining other distributions is left to the exercises.

Recall the definition of a continuous probability density function $f(x)$ defined over the interval $[\alpha, \beta]$: the area under this curve from a to b is the probability that x, chosen at random, lies in the interval $[a, b]$. Hence

$$\Pr(a \leq x \leq b) = \int_a^b f(x)\, dx$$

and of course

$$1 = \Pr(\alpha \leq x \leq \beta) = \int_\alpha^\beta f(x)\, dx.$$

A *uniform distribution* over the interval [0, 1] has as its density function a constant. We must have

$$1 = \Pr(0 \leq x \leq 1) = \int_0^1 c\, dx = c.$$

Thus the density function of this distribution is simply $f(x) = 1$.

Given a density function $f(x)$ defined over the interval $[\alpha, \beta]$, the mean or expected value, μ, is given by

$$\mu = \int_\alpha^\beta x f(x)\, dx,$$

and the expected fluctuation or standard deviation, σ, is given by

$$\sigma^2 = \int_\alpha^\beta (x - \mu)^2 f(x)\, dx.$$

Using these formulas, we find that $\mu = \frac{1}{2}$ and $\sigma = \sqrt{1/12}$ for the uniform distribution over [0, 1]. With this in mind we can explain how to approximate a *normal distribution* with mean μ and standard deviation σ, which is described by the density function

$$f(x) = \frac{1}{\sqrt{2\pi}\sigma} e^{-(x-\mu)^2/2\sigma^2}, \qquad -\infty \leq x \leq \infty.$$

This is a bell-shaped curve whose maximum is at $x = \mu$.

The key to such an approximation is a theorem known as the *central-limit theorem*, which was proved in its full strength by J. W. Lindeberg in the early 1920s, having been established earlier in a weaker form by A. Ljapunov and others. This famous theorem asserts that if $x_1, x_2, \ldots, x_n$ are independent random numbers chosen from a distribution $f(x)$ with mean μ and standard deviation σ, then the distribution of

$$y = \frac{\frac{1}{n}\sum_{i=1}^{n} x_i - \mu}{\sigma} \sqrt{n}$$

approaches the normal distribution with mean 0 and standard deviation 1. In other words, if we use uniformly randomly distributed real numbers x_i from [0, 1], in which case $\mu = \frac{1}{2}$ and $\sigma = \sqrt{1/12}$, then

$$y = \frac{\frac{1}{n}\sum_{i=1}^{n} x_i - \frac{1}{2}}{\sqrt{1/12}} \sqrt{n} = \left(\frac{1}{n}\sum_{i=1}^{n} x_i - \frac{1}{2}\right)\sqrt{12n}$$

is an approximation to the normal distribution with mean 0 and standard deviation 1, and this approximation becomes better and better as n gets larger and larger. If we want a normal distribution with mean M and standard deviation S, we can obtain it by using

$$y = M + S\left(\frac{1}{n}\sum_{i=1}^{n} x_i - \frac{1}{2}\right)\sqrt{12n}.$$

Since $n = 12$ is probably large enough for most applications, this simplifies to

$$y = M + S\left(\sum_{i=1}^{12} x_i - 6\right).$$

Using this formula, the generation of a single random number from the normal distribution will require us to generate 12 random numbers from the uniform distribution over [0, 1].

4.2. MONTE CARLO METHODS

The modern genesis of Monte Carlo methods is from research during World War II: scientists at Los Alamos Scientific Laboratory needed to know how far neutrons would travel through various materials. The answer to this question was crucial in the design of nuclear devices, but it was far beyond available analytical methods; experimental trial and error would have been too hazardous and time consuming. To solve this problem, Stanislaw Ulam, John von Neumann, and others resorted to a technique that had been known for almost a century, but which had been dismissed as a mere curiosity. The technique was based on the idea of using known physical properties and observed probabilities to predict, by means of a simulation, the outcome of experiments that could not be performed. At Los Alamos, this technique was given the code name Monte Carlo, since it was based on the same principles as intelligent gambling.

Generally speaking, a Monte Carlo method is any method that involves the use of random behavior to solve a problem. The methods are applied to broad classes of problems, deterministic and probabilistic. The former application may seem self-contradictory, but even in the study of completely deterministic problems, random behavior can be useful. Random variables are introduced, artificially, so that their expected value is the solution we want; we then simulate that random variable to approximate its expected value. Such methods can be used for a wide range of numerical mathematical problems, including the solution of differential equations, finding areas and volumes, inverting matrices, or calculating the value of π. Probabilistic applications concern the simulation of processes that *inherently* contain random variables, such as the simulation of telephone traffic through a switchboard or the servicing of cars at a garage. This section deals only with Monte Carlo methods applied to deterministic problems; simulation is discussed in Section 4.3.

Generally speaking, the application of Monte Carlo methods to deterministic problems has not been particularly successful. But for some deterministic problems, Monte Carlo methods provide the *only practical* means of solution. Perhaps the major disadvantage of such methods is their extremely slow rate of convergence. This convergence is governed by a square-root law which says that when there are n trials the error is proportional to $1/\sqrt{n}$. To clarify the meaning of this law consider, say, the calculation of π by the Monte Carlo method in the next section. The calculation of each decimal digit requires 100 times the work of the preceding one! Obviously, it is not practical to compute the result to more than two or three significant figures.

4.2.1. Buffon Needle Problem

Our first example of a Monte Carlo method is based on the observation by the eighteenth century French naturalist Comte de Buffon that the value of π can be computed by throwing needles onto a lined surface. Consider a plane ruled with parallel straight lines at distance $2d$ from each other. If a needle of length $2l$ is thrown at random onto the plane, what is the probability that the needle will intersect one of the straight lines? It turns out that the probability is given by a simple expression which involves π.

To analyze the situation, we shall describe the position of the needle at rest by the coordinates of its center, (x, y), and by the angle it forms with the x-axis, α. If we take the straight lines to be parallel to the x-axis, it is clear that the probability of intersection is independent of x and is a function only of y and α. Notice also that because of the symmetry of the problem we need only consider values of y modulo d (i.e., values of y between 0 and d), and values of α modulo $\pi/2$. We can now make precise the statement of the problem: an ordered pair of random values (y, α) is chosen so that y is uniformly distributed over $[0, d]$, α is uniformly distributed over $[0, \pi/2]$, and the two are independent of one another. In other words, every point in the rectangle shown in Figure 4.1 is equally likely to occur as the result of throwing the needle onto the plane. Now we must determine which points in that rectangle correspond to the needle intersecting a line. The analysis falls into two cases, when $d \geq l$, given next, and when $d < l$, which is left to Exercise 12.

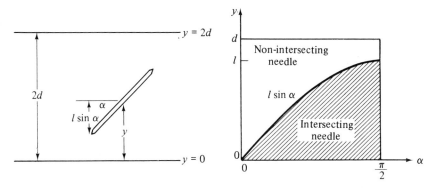

Fig. 4.1   The Buffon needle problem.

A needle with its center at (x, y), forming the angle α with the x-axis, will intersect a line if and only if $y \leq l \sin \alpha$. Only the points below the sine curve in Figure 4.1 satisfy this relation; the points above it do not. Thus the probability of an intersection is the ratio of the area under the curve, which is simply l, to the area of the rectangle, which is $\pi d/2$. So,

$$\text{Probability of intersection} = \frac{l}{\pi d/2} = \frac{2l}{\pi d}.$$

Now, the *law of large numbers* from probability theory, first stated by Jakob Bernoulli, says that with probability 1, the frequency of occurrence of an event in n trials converges to the probability of this event as n goes to infinity. From this law we conclude that we can estimate the probability of intersection as closely as desired by a long sequence of trials. Of course, as we indicated earlier, the square-root law tells us to expect that a very large number of trials would be needed to approximate π to only five decimal places. For this reason Buffon's needle problem is a poor method for calculating π; more efficient methods are described in Section 5.2.3.

What we have really done is to show how the area under a curve can be approximated using random numbers. Of course, in this case the curve was quite simple, but this method would have worked even if the curve were so irregular that it could not be integrated by other techniques. The next section shows how the Monte Carlo idea can be applied to multiple integration of arbitrary functions.

4.2.2. Areas and Volumes

In calculus one learns that, if F is the antiderivative of f (or, alternatively, $dF/dx = f$), then

$$\int_a^b f(x)\,dx = F(b) - F(a)$$

is the area under the curve $y = f(x)$ from $x = a$ to $x = b$. This rule, however, frequently does not provide a practical method for computing that area. There are several reasons for this; for example, we can integrate only the very simplest functions in closed form, or the function f may be given to us only in tabular form. Computationally, we can do better by recalling Riemann's original definition:

$$\int_a^b f(x)\,dx = \lim \sum_{k=1}^n f(x_k)(x_{k+1} - x_k),$$

where $a = x_1, x_2, \ldots, x_{n+1} = b$ is a partition of the interval (a, b) into subintervals, and the limit is taken over any sequence of partitions in which the length of the longest subinterval goes to zero. This provides a reasonable method of approximating $\int_a^b f(x)\,dx$ to any desired accuracy.

Although numerical integration is generally an efficient and reliable process, there is a major class of problems in which it presents serious difficulties. This is the case of multiple integrals over complicated regions in high-dimensional spaces. Methods that use successively finer subdivisions on the line become increasingly difficult in high-dimensional spaces. Even if the boundaries of the region of integration are simple (and they usually are not), the number of points at which the function to be integrated needs to be evaluated grows enormously; a 10-point subdivision of an interval, which may only give a crude approximation, generates 1000 points in three dimensions and 1 million in six dimensions. It is precisely multiple integrals over complicated regions that are the motivation for Monte Carlo methods of integration.

In describing the basic parts of the method, we shall assume the special case of computing the volume of an arbitrary solid shape. The generalization of this technique to the case $\iint \ldots \int f(x, y, \ldots, z)\, dx\, dy \ldots dz$ is straightforward. Suppose that we are given the solid region R by having its boundaries defined as relations between three variables; we want to compute the volume of the region R. We begin by considering a region Q, containing R, whose volume is easy to compute; usually Q is a cube. Now, using a random-number generator, choose points

$$p_1, p_2, p_3, \ldots, p_k, \ldots$$

at random inside the region Q. For each point p_i, check whether or not it is in the region R, and keep a running count of the number of points that do lie in R. Let this count be denoted by $V(k)$.

We now appeal to the law of large numbers, as before, to conclude that

$$\lim_{k \to \infty} \frac{V(k)}{k} = \frac{\text{volume of } R}{\text{volume of } Q}.$$

Hence for a large enough value of k we have

$$\text{volume of } R \approx \frac{V(k)}{k} \cdot \text{volume of } Q.$$

But, how large must k be? This question has been studied extensively, and a discussion of the answer requires more of the theory than we have been able to present. We can, however, make somewhat more precise the square-root law to which we alluded earlier. Let

$$r = \frac{\text{volume of } R}{\text{volume of } Q},$$

then it can be shown that for all $\epsilon > 0$

$$\Pr\left(\frac{V(k)}{k} - r \leq \sqrt{\frac{r(1 - r)}{\epsilon k}}\right) = 1 - \epsilon.$$

This relation allows us to estimate the number of points required for a given accuracy.

To illustrate this Monte Carlo method, let us use the random-number generator of Section 4.1.1 to compute the volumes of the following three regions, all of which are contained in the unit cube, $0 \leq x, y, z \leq 1$:

1. An inscribed sphere of unit diameter.
2. The cube given by $\frac{1}{4} \leq x, y, z \leq \frac{3}{4}$.
3. The set R of all points that satisfy the inequalities

$$x^2 \sin(10\pi y) \geq \sqrt{z},$$

$$xyz \geq 0.001,$$

$$0 \leq x, y, z \leq 1.$$

Successive points p_k were chosen for $k = 1, 2, \ldots, 600$ by taking

$$p_k = (r_{3k-2}, r_{3k-1}, r_{3k}),$$

where $r_1, r_2, \ldots$ is the sequence of random numbers produced by the generator of Section 4.1.1, appropriately scaled. One can see from the calculations in the table for the sphere and cube, whose exact volume we know, that our calculation of the volume of R will not be accurate to more than one digit, since we only used 600 random points. However, we cannot increase the number of points substantially with our random-number generator because its period is only 2048, and we would simply be repeating points that had already been generated.

Number of Points	Sphere	Cube	R
100	0.540	0.110	0.040
200	0.465	0.095	0.055
300	0.507	0.107	0.057
400	0.512	0.102	0.048
500	0.510	0.112	0.046
600	0.513	0.110	0.052
Exact volumes	0.525	0.125	?

What is the advantage of using a random selection of points rather than a regular grid of points? If we interpret "at random" as meaning "lack of any pattern," it is clear that a random choice is safer than a regular grid, since a regularity in the points could interact with a pattern in the boundaries of the region whose volume we are calculating, producing erroneous results. In connection with this point we should mention that it is prudent to repeat any Monte Carlo calculation using a different random-number generator; this removes the possibility of a hidden interaction between the random-number generator and the boundaries of the region. Having done this, if the two results are in close agreement, we can be reasonably certain that our answer is free from errors due to pattern interaction.

4.2.3. Random Walks and Potential Theory

In this section we shall describe a Monte Carlo method that is quite different from those considered previously. Imagine a drunken man who is trying to find his way home from a bar. He is, in fact, so drunk that at each intersection he proceeds randomly in some direction; upon reaching the next intersection he again chooses a direction at random. Our drunken friend is going on what is usually called a *random walk*: starting at the origin at time $t = 0$, take a step of fixed length in a direction chosen at random according to some probability density function (which is independent of the current position). Taking this step requires one unit of time, and the process is repeated for $t = 1, 2, 3, \ldots$. An interesting example of a random walk is Brownian motion. At the beginning of the nineteenth century Robert Brown, a Scottish botanist, observed that small particles suspended in water performed a rapid oscillatory motion. This Brownian motion was later explained as the result of impacts of molecules, which, according to the kinetic theory of heat, are in constant motion.

For simplicity, we shall assume a one-dimensional space, but the derivation below does not really depend on this assumption. Let $p(x, t)$ be the probability of finding a particle at position x at time t, assuming that it had been observed at the origin $x = 0$ at time $t = 0$. In 1905, Einstein showed that the function $p(x, t)$ should be a normal distribution

$$p(x, t) = \frac{b}{\sqrt{t}} e^{-cx^2/t}, \qquad t > 0, \qquad -\infty \leqq x \leqq \infty,$$

where b and c are some constants.

Three observations about $p(x, t)$ should be made. First, since the probability of finding the particle somewhere must be 1, we have

$$\int_{-\infty}^{\infty} p(x, t)\, dx = 1.$$

Second, at any time $t > 0$, $p(x, t)$ is a bell-shaped curve with its maximum at $x = 0$; this means that the single most likely spot to find the particle is always where it was last observed. Third, as t increases, the height of $p(x, t)$ at the maximum $x = 0$ decreases, and since the area under the curve must be 1, $p(x, t)$ must increase elsewhere, that is, for large values of $|x|$. This means that the probability of finding the particle far from the origin increases as t increases.

What is the average distance, $A(t)$, of a particle from the origin at time t? We have

$$A(t) = \int_{-\infty}^{\infty} |x|\, p(x, t)\, dx.$$

Since the integrand is symmetric about $x = 0$, this becomes

$$A(t) = 2 \int_0^\infty xp(x, t)\, dx = 2 \int_0^\infty x \frac{b}{\sqrt{t}} e^{-cx^2/t}\, dx.$$

Substituting $y = cx^2/t$, we obtain

$$A(t) = C\sqrt{t}$$

for some constant C. This is another square-root law, and it tells us that the average distance that a particle subjected to a random walk is displaced from the origin is proportional to the square root of the duration of the walk.

We shall now show how the idea of a random walk can be used to solve, computationally, problems whose analytical solutions are difficult or impossible to find. Such problems were among the first to be solved with Monte Carlo methods. The problems we want to solve using random walks come from the branch of applied mathematics known as *potential theory*, which studies what are known as *harmonic functions*. These functions arise whenever we have a homogeneous medium in a state of equilibrium. Suppose that we have a sheet of metal whose boundaries are subjected to enough heat to keep them at a fixed temperature. After a while, the sheet of metal will reach equilibrium: the temperature at the interior points on the sheet will stop fluctuating, and at that time the temperature T at each of the interior points (x, y) will be a function of x and y; more concisely, it will be $T(x, y)$. Now it is clear from our intuition, and it can be rigorously proved mathematically, that the function T is continuous and has the *mean-value property* which says that if (x_0, y_0) is an interior point on the sheet of metal and Γ is any circle centered at (x_0, y_0), lying entirely within the region of the sheet of metal, then $T(x_0, y_0)$ is the average of all values of $T(x, y)$ for (x, y) in Γ. A function with these properties is called a *harmonic function*. Such functions arise in physical situations such as heat flow and elastic membranes.

The connection between harmonic functions and random walks is the surprising fact, first shown by the Japanese mathematician Shizuo Kakutani, that random walks can be used to solve the well-known Dirichlet problem. Suppose, as in the previous example, that a surface is at a thermal equilibrium, with its temperatures specified at the boundaries. The Dirichlet problem is to compute the temperature at interior points, given the values on the boundaries. This problem is amenable to analytic techniques when the boundary is simple, say a rectangle or circle; but when the boundary is "messy," one must resort to the Monte Carlo method about to be described.

Suppose that we want to find the value of T at a given point (x_0, y_0). Using (x_0, y_0) as the starting point, begin a random walk. By the square-root law, this walk will hit the boundary with probability 1, and at that time we record the temperature at that boundary point, and start a new random walk at (x_0, y_0), and repeat the process. Eventually, we will have recorded a number of temperatures at the boundary points we encountered on these random walks. Now, as the number of random walks goes to infinity, the average of these recorded temperatures goes to $T(x_0, y_0)$. In other words, the expected outcome of a random walk starting at (x_0, y_0) is $T(x_0, y_0)$.

Intuitively, this method makes sense, for if we start near a hot region of the boundary, it is likely that we shall terminate at a boundary point of high temperature, and if we start near a cool region of the boundary, it is likely that we shall terminate at a boundary point of low temperature. Thus the regions near hot boundaries will be hot, and regions near cool boundaries will be cool. By neglecting some mathematical fine points, we can back up this intuition with a proof that the expected outcome of a random walk starting at (x_0, y_0) is the equilibrium temperature of that point. For convenience, let $E(x, y)$ be the expected outcome of a random walk starting from the point (x, y). If (x, y) is on the boundary, the walk terminates without taking a step, and thus $E(x, y) = T(x, y)$ for all boundary points. To show that these functions are equal on interior points, it is sufficient to show that E is harmonic, since T is, by definition, harmonic, and a harmonic function with given boundary conditions is unique; see Exercise 18.

The continuity of E is clear: if the starting point is shifted slightly, the random walk is only slightly affected and the probability of encountering a given boundary point is only slightly changed. We must also show that E satisfies the mean-value property. Let Γ be a small circle around some interior point (x_0, y_0); we must show that $E(x_0, y_0)$ is equal to the average of the values of $E(x, y)$, where (x, y) is an arbitrary point on Γ. Pick a point (u, v) on Γ and consider only those random walks starting at (x_0, y_0) that go through (u, v). If we take the average of the outcomes of all such random walks over all the points on Γ, then this is clearly equal to $E(x_0, y_0)$, since the average of the subaverages is equal to the average. Now, what are we taking the average of? A random walk depends only on the *current position* of the particle, and not on its history; thus the expected value of the random walk starting at (x_0, y_0) and going through (u, v) is the same as the expected value of the random walk starting at (u, v). So we are taking the average of $E(u, v)$ for all points (u, v) on Γ, and we conclude that this equals $E(x_0, y_0)$; so E is indeed harmonic.

The most noteworthy feature of the Monte Carlo method we have described is that it can be used no matter how irregular the boundary. Boundaries that are so nasty as to defy analytical techniques and nonprobabilistic numerical techniques can be easily handled by this Monte Carlo method.

4.3. SIMULATION

There is really no clear dividing line between the Monte Carlo computations treated in the previous section and the simulation problems discussed in this section. Typically, the problems previously considered had randomness introduced artificially; in the problems here, if randomness is present, then it is inherent. The term *simulation* is used to describe experimentation with models of systems in order to determine their behavior; the usefulness of simulation as a technique lies in the fact that the system need not actually be constructed to be simulated. For example, suppose that we want to construct a network of roads which form interconnections between cities, and we want to determine the most efficient manner to connect various cities. It is certainly not feasible to build a system of roads, see how it works out, and then modify it if need be. But we can develop a computer program to simulate the effect of various road systems and use that simulation to tell us, *before* we build the roads, what the modifications should be. Simulation provides us with a tool to forecast the effects of alterations without actually having to make them; thus undesirable features of the system can be eliminated before even a proto-type is built.

Simulation need not involve random events. For example, a computer manufacturer might design a hypothetical computer and write a program for an existing computer, which could run programs written for the hypothetical computer. In this way programmers could develop software for the nonexist-ent computer and then suggest design changes based on their experience. Alternatively, when a new computer is put onto the market, many people have a financial interest in the programs for their current computers. A pro-gram for the new computer that simulates the old one makes those programs usable until they can be reprogrammed for the new machine.

As examples of typical simulation problems, we shall consider two models of traffic flow: one of the intersection of two roads by means of a traffic circle or "roundabout," and the other of a single-lane no-passing stretch of road; the former involves random events and the latter does not. Although both examples deal with the simulation of traffic, we should point out that simulation methods are far more widely applicable. Such techniques have been successfully used to study many systems in which the various interac-tions are much too complicated to be treated analytically; examples from queuing theory include the simulation of businesses in which customers arrive and are serviced with certain frequencies, airplane traffic flow, and telephone exchanges.

Both models we consider are greatly oversimplified, and so we cannot expect to draw valid, practical conclusions from them; they are intended only to illustrate some of the principles and techniques used in the design of simulation models. We leave as an exercise the problem of making these two models more realistic.

4.3.1. Circle or "Roundabout"

We shall consider the traffic pattern illustrated in Figure 4.2. Roads from the east, north, west, and south lead into and out of a traffic circle. Our

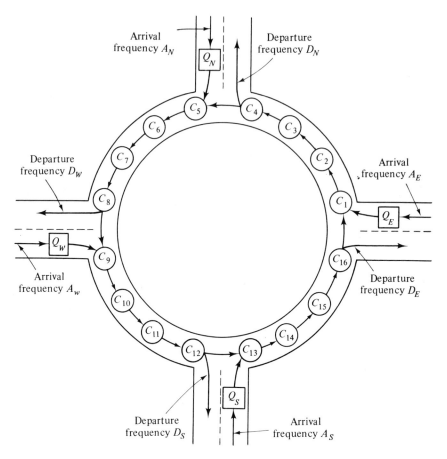

Fig. 4.2 The traffic circle or "roundabout."

model will be discrete in both space and time, since we shall assume that there are a fixed number of positions (16) in the circle which may be occupied by cars, and that motion occurs only at times $t = 1, 2, 3, \ldots$, when a car advances to the position immediately ahead of its current position. We also assume that a car in the circle has the right of way, and hence an arriving car must wait to enter the circle until the position to its left is empty (for example, position 16 for a car arriving from the east). This may cause queues to build up, and we denote by Q_E, Q_N, Q_W, and Q_S the number of cars waiting in a queue on each of the four roads.

For each simulation run we shall specify fixed arrival frequencies A_E, A_N, A_W, and A_S on each of the four roads and fixed probabilities P_R, P_S, P_L, and P_U that a car arriving at the circle will turn right, go straight, turn left, or make a U-turn, respectively. Thus $A_E = \frac{1}{3}$ means that, on the average, at one out of every three time instants, a car will arrive from the east and so Q_E will be incremented by 1. Similarly, $P_R = P_S = P_L = 0.3$ and $P_U = 0.1$ mean that, on the average, three out of ten cars drive through one quarter of the circle, three out of ten drive through one half of the circle, three out of ten drive through three quarters of the circle, and one out of ten drives through the entire circle.

The result of each simulation will be the observed departure frequencies D_E, D_N, D_W, and D_S for each of the four roads. The sum of the departure frequencies $D = D_E + D_N + D_W + D_S$ measures the flow of traffic through the circle. The circle is saturated when D becomes less than the sum of the arrival frequencies $A = A_E + A_N + A_W + A_S$, and in this case the arrival queues grow without bound.

One of the most important questions about this model is when does saturation occur, as a function of the arrival frequencies and the turning probabilities? The graph shown in Figure 4.3 shows D as a function of A in the special case that each of the arrival frequencies is equal to $A/4$, and for the turning probabilities $P_R = P_S = P_L = 0.3$, $P_U = 0.1$. As is expected, D equals A for small values of A; that is, as many cars depart as arrive. However, the circle cannot accommodate a traffic volume beyond about $A = 2$, corresponding to arrival frequencies on each road of about 0.5.

Of all the simplifications on which this model is based (e.g., all cars have the same constant speed, or a car stopped in the queue accelerates instantaneously) perhaps the most serious is that we have ignored any interaction be-

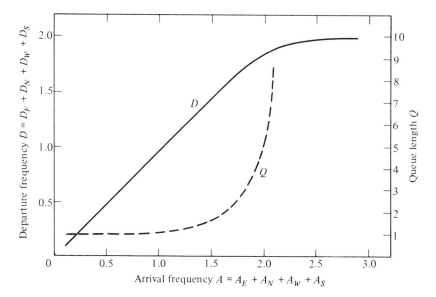

Fig. 4.3   Flow through the traffic circle.

tween cars. We know from experience that as traffic becomes denser the behavior of a car is determined more and more by the surrounding cars and less and less by the driver's wishes. Our next model is intended to study this interaction between cars.

4.3.2. Single-Lane Traffic

Let us consider the flow of traffic along a single-lane no-passing stretch of road. The model we shall use, called the *follow-the-leader theory*, is based on the observation that whenever a driver comes close enough (between 20 and 300 feet, depending on the speed of the car) to the car ahead of him he begins to interact with it. In a long line of cars, this interaction can cause disturbances which propagate like shock waves, and which can be amplified to the point that they cause a collision. The question we want to answer is what traffic flow can the road accommodate, as a function of car density.

There are, of course, many assumptions and definitions that must be made before we can perform the simulation. But whatever assumptions and definitions we make, we must have as a qualitative result from our theory that which we know from experience: as long as car interactions occur rarely the traffic flow will increase as the car density increases. At some point, however, it will begin to decrease until, at saturation, a bumper-to-bumper standstill results.

Cars on a highway tend to travel in clusters rather than being uniformly spaced along the road. In addition, a particular driver's actions can be assumed to be influenced more by the car immediately ahead of him than by any other car in the cluster. In addition to this assumption we shall make two others. First, a driver will always try to match his speed with the car ahead of him, and the force of his actions to match that speed is greater as the difference between their speeds increases and is smaller as the distance between their cars decreases. In particular we shall express this assumption as

$$\text{change in acceleration} = \text{constant} \frac{\text{difference in speeds}}{\text{separating distance}}.$$

The proportionality constant in this equation can be regarded as an indicator of a driver's habits; a low coefficient means the driver has a light touch on the pedals; a high coefficient means that he is relatively "heavy footed." To achieve realistic results, it has been found that rather tight upper and lower bounds on this coefficient must be imposed. The second assumption, designed to simplify the model, states that all the drivers have identical reaction times; this reaction time is the time increment between steps of the simulation.

We have n cars, numbered 1 to n arranged so that for $1 < i \leq n$, car i follows car $i - 1$. The lead car (number 1) can be controlled independently of the others as a special case, or it can be made to follow the nth car, as if the cars were on a closed tack. To start the simulation, we assign to each car $i, i = 1, \ldots, n$, the following characteristics:

$c_i =$ driver constant (this remains fixed throughout the simulation),

$x_i =$ initial position, or the distance from a fixed reference point,

$x_i' =$ initial velocity (in distance per time step),

$x_i'' =$ initial acceleration (in distance per time unit squared).

Then, for each time step we calculate for $i = 2, \ldots, n,$

$$x_i'' = c_i \frac{x_{i-1}' - x_i'}{x_{i-1} - x_i},$$
$$x_i' = x_i' + x_i'',$$
$$x_i = x_i + x_i'.$$

The calculations made for the first car ($i = 1$) depend on what we want to simulate. A circular track would be simulated by

$$x_1'' = c_1 = \frac{x_n' - x_1'}{x_n - x_1},$$
$$x_1' = x_1' + x_1'',$$
$$x_1 = x_1 + x_1'.$$

To prevent cars from backing up we set, x_i' to zero when it becomes negative.

The results of a simulation of this model are shown in Figure 4.4. For this run, there were 25 cars each of which had drivers with constants of 50, initial velocities of 80 feet/second, and initial accelerations of 0 feet/second/second; the cars were positioned 200 feet apart on a straight road (i.e., the lead car's behavior was independent of the others). The lead car was put into a controlled stop, decelerating at 5 feet/second/second from 80 feet/second to 0 feet/second. The graph shows a plot of distance versus time for the first 10 cars, and the shock wave effect is clearly demonstrated: in order to avoid a rear-end collision, the cars later in the string decelerate far more rapidly than the first car.

In both of our examples, the simulation process did not involve interaction with any *real traffic* at all. However, sometimes simulation is carried out in order to control the very process being simulated. For example, we might have automatic sensors on the road to detect the number of cars and their projected arrival times as they approach a critical stretch of road. If the simulation can be done in *real time*, that is, if it can be used to predict the traffic

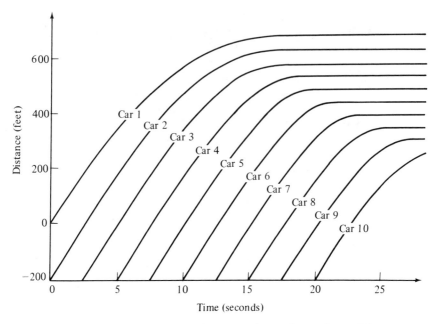

Fig. 4.4 The shock wave effect in the interacting traffic model. Initial velocity 80 ft/s, initial spacing 200 ft apart, initial acceleration 0 ft/s². Car 1 was put into a controlled stop by having it decelerate at a constant 5 ft/s².

situation in advance of its occurrence, then the controller (either human or automatic) may be able to take preventive measures to avoid undesirable situations.

4.4. REMARKS AND REFERENCES

The general idea of simulation with the aid of stochastic processes is quite old, going back almost to the beginnings of probability; however a systematic development of the idea of using random sampling to approximate distribution functions is due, apparently, to Student in

STUDENT. "The Probable Error of a Mean," *Biometrika*, 6 (1908), 1–25.

In those days random numbers were generated by drawing cards from a shuffled deck, rolling dice, or some other similar process. This is quite tedious, particularly when many numbers are needed. So, at the suggestion of Karl Pearson, L. H. C. Tippett prepared a list of "random" digits selected from census reports! This table of over 40,000 digits was published as

TIPPETT, L. H. C. "Random Sampling Numbers," *Tracts for Computers*, No. 15, Cambridge University Press, New York, 1927.

In 1939 M. G. Kendall and B. Babington-Smith published the first table of random digits generated by a machine. In addition to describing their machine, their papers

KENDALL, M. G., and B. BABINGTON-SMITH. "Randomness and Random Sampling Numbers," *J. Roy. Stat. Soc.*, *101* (1938), 147–166,

and

————. "Second Paper on Random Sampling Numbers, *J. Roy. Stat. Soc.*, *Suppl.* 6 (1939), 51–61,

also introduced the frequency, poker, and gap tests for random numbers. The most extensive table of machine-generated random numbers can be found in

Rand Corporation. *A Million Random Digits with 100,000 Normal Deviates*, Free Press, div. of Macmillan, New York, 1955.

What we have called "perfect infinite random sequences" arise in number theory as so-called *normal numbers*. A complete development of this topic is given in Chapter 8 of

NIVEN, I. *Irrational Numbers*, No. 11 in the series of Carus Mathematical Monographs, distributed by Wiley, New York, 1956.

The use of arithmetic schemes to generate "pseudorandom" numbers is less than 30 years old, and it is due to J. von Neumann and N. Metropolis in 1946. They suggested the well-known *center-squaring* technique in which the next number in the sequence is obtained by taking the middle digits of the square of the previous number. This method has never been successfully analyzed and it does not always produce satisfactory results. It was D. H. Lehmer who first proposed the idea of the linear congruential method in 1949. The random-number generator given in Section 4.1.1 is of this type, and it is described in

KRUSKAL, J. B. "An Extremely Portable Random Number Generator," *Comm. ACM*, *12* (1969), 93–94.

An excellent review article with a comprehensive bibliography (now slightly dated) is

DOBELL, A. R., and T. E. HULL. "Random Number Generators," *SIAM Rev.*, *4* (1962), 230–254.

The third chapter of

KNUTH, D. E. *The Art of Computer Programming*, Vol. 2, Seminumerical Algorithms, Addison-Wesley, Reading, Mass., 1969 contains a complete description of the current state of the art in the generation and testing of random numbers.

The first Monte Carlo method published is probably that of Comte de Buffon (1707–1788), described in Section 4.2.1. It appeared in 1777 in his "Essai d'arithmétique morale," which was a part of the fourth volume of his major work *Histoire naturelle*. The best approximation to π obtained by Buffon's method was made in 1901 by the Italian Lazzerini. From only 3408 tosses of the needle he claimed to have found π correct to six decimal places; his results are regarded with some suspicion!

The first major Monte Carlo calculations were connected with the design of atomic weapons at Los Alamos during World War II. J. von Neumann and S. Ulam are usually credited with the major ideas that got Monte Carlo started. Much of this early work is described in

> MEYER, H. A. (ed.). *Symposium on Monte Carlo Methods*, Wiley, New York, 1956.

A more recent introduction to Monte Carlo techniques is given in

> HAMMERSLEY, J. M., and D. C. HANDSCOMB. *Monte Carlo Methods*, Methuen, London, and Wiley, New York, 1964.

The techniques by which harmonic equations can be solved using a Monte Carlo method are described in

> HERSH, R., and R. J. GRIEGO. "Brownian Motion and Potential Theory," *Sci. Amer.*, *220*, No. 3 (Mar. 1969), 66–74.

Simulation is an area of computer application that has developed very rapidly. Surveys of the area and many references to specialized literature can be found in

> MARTIN, F. F. *Computer Modeling and Simulation*, Wiley, New York, 1968,

> MAISEL, H., and G. GNUGNOLI. *Simulation of Discrete Stochastic Systems*, Science Research Associates, Chicago, 1972.

> GORDON, G. *System Simulation*, Prentice-Hall, Englewood Cliffs, New Jersey, 1969.

We have chosen to illustrate simulation techniques with traffic problems because they are familiar to everybody and of obvious importance in today's society; and in the past 10 years there has emerged a mathematical theory of traffic flow. See, for example,

> ASHTON, W. D. *The Theory of Road Traffic Flow*, Methuen, London, and Wiley, New York, 1966.

The follow-the-leader theory on which our second example is based is described in

HERMAN, R., and K. GARDELO. "Vehicular Traffic Flow," *Sci.
Amer.*, *209*, No. 6 (Dec. 1963), 35–45.

In 1960, Toronto became the first major city to put traffic signals under computer control. The status of such traffic systems, as of 1968, is described in

FRIEDLANDER, G. G. "Computer-Controlled Vehicular Traffic,"
IEEE Spectrum, *6* (1969), 30–43.

4.5. EXERCISES

1. The random number generator given in Section 4.1.1 is an example of a linear congruential method; that is,

$$r_{n+1} = ar_n + b \ (\text{mod} \ m)$$

for certain constants a, b, and m. Discuss the effects of the choice of these three constants on the efficiency of the generator (how difficult is it to compute the next number?) and on the period of the generator (can the maximum possible period of m be realized; if so, how?).

2. The standard deviation, σ, of a random variable x with mean μ is the square root of the expected value of the square of the difference between x and μ:

$$\sigma = \sqrt{E((x - \mu)^2)}.$$

Show that the standard deviation of $O(k)$, the occupancy function from Section 4.1.1, is proportional to $\sqrt{k}$.

3. Suppose that you are given approximate measurements x_i to n numbers X_i, $i = 1, 2, \ldots, n$. The accuracy of the measurements is to within 0.01; that is, $|x_i - X_i| \leq 0.01$; assume that the values of $x_i - X_i$ are uniformly distributed in the interval $(-0.01, 0.01)$. The absolute value of the error in $\sum x_i$, as an approximation to $\sum X_i$, is certainly less than $0.01n$. Find the expected error and the standard deviation of that error.

4. Explain why the chi-square statistic for Table 4.1 gives percentages that get closer and closer to 100 per cent as k goes from 100 to 1000.

5. The *gap test* examines the lengths of "gaps" between occurrences of random numbers in some range. Suppose that our random numbers are uniformly distributed in the interval [0, 1]. Let $0 \leq \alpha < \beta \leq 1$. The test is applied by recording the lengths of sequences of random numbers that are outside the range $[\alpha, \beta]$ and comparing the results with the expected distribution by the chi-square test. Compute the probabilities of gaps of length $0, 1, 2, \ldots,$ $n - 1$, and gaps longer than $n - 1$. Discuss the two special cases of this test when $\alpha = 0$, $\beta = \frac{1}{2}$ and $\alpha = \frac{1}{2}$, $\beta = 1$ (runs above and below the mean, respectively). Discuss the relation between this test and the frequency test. Apply this test to the random-number generator of Section 4.1.1.

6. Apply the poker test to the random-number generator of Section 4.1.1.

7. Describe and use on some sample random numbers the following simplification of the poker test: instead of counting poker hands, count the number of distinct values in a set of five random integers.

8. Find a method for generating random permutations from random numbers. Each permutation should have an equal probability of occurrence.

9. How can random integers in a Poisson distribution with mean μ be generated using uniformly distributed random numbers in $[0, 1]$? *Hint:* Let $r_1, r_2, \ldots,$ r_n be numbers in $[0, 1]$ generated by the random generator. Prove that the probability of the inequalities $r_1 r_2 \ldots r_n \geq x$ and $r_1 r_2 \ldots r_{n+1} < x$ holding simultaneously is

$$\frac{x \left(\ln \frac{1}{x} \right)^n}{n!}, \qquad 0 < x \leq 1.$$

This follows by induction on n and the integration of

$$\frac{1}{(n-1)!} \int_x^1 \frac{x}{r} \left(\ln \frac{r}{x} \right)^{n-1} dr.$$

Choose $x = e^{-\mu}$ and then show that n is distributed according to the Poisson distribution,

$$\Pr(n = i) = \frac{e^{-\mu} \mu^i}{i!}.$$

10. Prove that the random variable x, generated as follows, is normally distributed with mean 0 and standard deviation 1: generate a pair (r_1, r_2) of uniformly distributed values in $[0, 1]$ such that $r_1^2 + r_2^2 < 1$; you may have to generate several pairs before getting one that satisfies this condition. Then compute

$$x = r_1 \sqrt{\frac{-2 \ln (r_1^2 + r_2^2)}{r_1^2 + r_2^2}}.$$

How many pairs (r_1, r_2), on the average, must be generated before you find one that satisfies the condition?

11. The exponential distribution with mean μ has density function

$$F(x) = 1 - e^{-x/\mu}.$$

Show that if r is uniformly distributed in $[0, 1]$ then $x = -\ln r$ is exponentially distributed with mean 1.

12. Complete the analysis of the Buffon needle problem by considering the case $d < l$.

13. In Lewis Carroll's famous *Pillow Problems*, problem 58 is to compute the

probability that three points chosen at random on an infinite plane form an obtuse triangle. Estimate this probability by performing experiments on the computer; the actual value is

$$\frac{3}{8 - (6\sqrt{3}/\pi)} \approx 0.6394.$$

14. By performing experiments on the computer, estimate the probability that three points chosen at random on the boundaries of the unit square form an obtuse triangle. The actual value is $\frac{19}{36} \approx 0.528$.

15. Approximate the value of π by using a Monte Carlo technique to compute

$$\int_0^2 \sqrt{4 - x^2}\, dx.$$

16. Consider the area A in the xy-plane defined by the inequalities

$$0 \leq x \leq 2,$$
$$\frac{x^2}{4} \leq y \leq \frac{3 \sin(32\pi x + \pi) + 5}{8}.$$

We want to compute the area A by a Monte Carlo calculation using the random-number generator of Section 4.1.1 to obtain points in the rectangle

$$0 \leq x \leq 2,$$
$$0 \leq y \leq 1.$$

This can be done by separating the 13-bit number r_n into its 7 high and 6 lower-order bits and scaling each of these numbers to obtain points in the desired rectangle. What will happen in this computation? What would be the effect of replacing sin by cos in the definition of A? Using the same random-number generator, modify the construction of the points (x_n, y_n) so as to obtain the area A more accurately.

17. Contrive an example in which two different random-number generators lead to significantly different results in estimating the area of some region in the xy-plane. Neither of the random-number generators should be "obviously" bad; the difficulty should stem from interaction between one of the random-number generators and the boundary of the region. (*Hint:* Use the random-number generator of Section 4.1.1 on a region whose boundaries are biased with respect to integers of the form $4k + 1$.)

18. Show that if $f(x, y)$ and $g(x, y)$ are harmonic functions then $f(x, y) - g(x, y)$ is also harmonic. Then, by showing that a harmonic function must achieve its maximum and minimum on the boundary, demonstrate that a harmonic function with specified boundary conditions is unique.

19. An interesting question about the traffic circle of Section 4.3.1 is how big are

the fluctuations in the size of the queue for traffic volumes below the saturation point. Under the same assumptions that $A_E = A_N = A_W = A_S = A/4$, $P_L = P_S = P_R = 0.3$, and $P_U = 0.1$, run simulations to obtain a graph of the average queue length as a function of $A \leq 2$.

20. Design and program a simulation model for a gas station with two pumps. Assume that customers arrive at an average rate of 20 per hour and that it takes exactly 5 minutes to service a customer. If both pumps are busy, the probability of a customer waiting is inversely proportional to the number of cars in the shorter of the two lines waiting at the pumps; if the customer does decide to wait, he goes to the back of the shorter line. Simulate the gas station for a 2-hour period with an increment of 1 minute (in any given minute, the probability of a customer arriving is $\frac{1}{3}$). Determine the average time a car spends in the gas station, the number of cars per hour that do not wait, and the average queue length at the pumps. Make your model more realistic by assuming that service time is not a constant, but rather is a random variable, uniformly distributed between 1 and 5 minutes. Be certain to perform your simulations several times with different random sequences to assure reliable results.

21. A press conference is attended by K reporters in a hall equipped with K telephones. At the end of the press conference, each reporter, eager to file his story rushes for a phone. Suppose that all K phones are equally accessible to each reporter and that he chooses a phone at random, independently of the other reporters. If several reporters head for the same phone, only one gets it, and the remaining reporters are forced to choose again among the remaining phones until each reporter is matched to a phone. Design and program a simulation model of this process that could be used to study such questions as: How many tries (as a function of K) must be made on the average before each reporter is matched to a phone? How would the introduction of i extra phones, $i = 1, 2, 3, \ldots$, affect the matching time?

22. Make the simulation model of Section 4.3.2 more realistic by whatever changes you desire. Implement that modified model and compare the results with those shown in Figure 4.4.

23. There are n bugs on the floor at points (x_i, y_i) for $i = 1$ to n. Each bug chases the next one: bug 1 chases 2, 2 chases 3, $\ldots$, $n - 1$ chases n, and n chases 1.

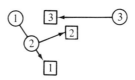

They each head directly toward the bug they are chasing and they all move *simultaneously*. Each time interval every bug moves one unit in the direction in which the bug he is pursuing was at the end of the *previous* time interval. So, if we have the configuration illustrated, where circles represent the state at time t, then squares would be the state at time $t + 1$. Develop a mathematical model of this problem, then write a program to simulate the chase using your model. Your program should read in the number of bugs, their initial positions, and for how many time intervals the chase is to be simulated. Run several cases, say four and six bugs, each with varied starting positions. Analytically, what can you say about the problem for two bugs? For three bugs? For four bugs? For n bugs?

5 COMPUTING WITH NUMBERS

The traditional meaning of computation always implied the manipulation of numbers, and the conventional view of computers frequently is that they are "number crunchers." Recently, attitudes toward computers have been changing, and, as a consequence, the usage of the word "compute" has been changing. It is not unusual to say that a machine computes, for example, with pictures, written sentences, or sounds when it processes photographs, when it analyzes, translates, or generates text, or when it recognizes spoken words. Most of this book discusses the techniques employed when a computer is used to process structures more complicated than numbers, such as algebraic expressions or graphs.

It is, of course, possible to view any activity performed by a digital computer as a computation on numbers, say, as a transformation of a single integer (possibly a very large one) into another. In most cases, however, there is nothing to be gained by taking that view; in fact, much intuitive meaning may be lost. In the sense of the intuitive meaning of a computation, the relative amount of numerical computation is decreasing while computer uses such as managing files of information, controlling assembly lines, editing text, or acting as a message switching center are becoming increasingly popular.

Numerical computation, however, still remains important. Foremost in importance is the field of numerical analysis, which deals with the problem of how computations on real numbers can be approximated by computations in the discrete number systems of digital computers. We mention some of the problems that arise in Section 5.1, but the interested reader should study any of a large number of texts in this field.

In Section 5.2 the computation of mathematical constants is described.

Here we examine some of the techniques used when numbers have to be computed to extremely high precision, such as to one million decimal places.

Finally, in Section 5.3 we give some examples of how number theoretic problems have been approached with a computer. In this class of problems it is frequently the case that a computer could disprove a conjecture by finding a counterexample, or it can be useful indirectly in proving a conjecture. This gives rise to interesting man–machine approaches, where the computer is used to provide data, which may lead to a new conjecture that must be tested by generating more data, and so on. This process is shown as a case study in attacking a difficult number theoretic problem in Section 5.3.3.

5.1. COMPUTER ARITHMETIC AND REAL NUMBERS

During the last three decades there has been a great emphasis placed on the study of algorithms that approximate, to any desired degree of accuracy, the solutions to various problems. This field, called numerical analysis, owes the interest in it to the current widespread use of digital computers. Old methods used in doing calculations by hand or by a manually operated calculator are not generally useful for computers, since they involve heuristic tricks and fail to take advantage of the great speed with which computers can do arithmetic. A fast modern computer can perform millions of arithmetic operations per second; few people do as much numerical computation in a lifetime as a computer can do in one minute. Since, as we shall see, the computer representation of real numbers is inexact, much attention must be focused on the estimation and control of errors that can propagate during lengthy computations. Anyone who uses a computer for numerical work must understand the basic principles of numerical analysis in order to know what relation there is between the numbers printed out after a long computation and the true answers to the problem to be solved.

To discuss this provocative subject, we shall look at the representation of real numbers in computers by means of the usual floating-point notation. We shall then discuss two of the most important aspects of numerical computation: convergence and stability.

5.1.1. Floating-Point Representation and Roundoff Errors

In the course of a computation, numbers can arise that are astronomically large, like

$$5980000000000000000000000000000,$$

which is the mass of the earth in grams, or microscopically small, like the mass of an electron in grams:

$$0.0000000000000000000000000009 1091.$$

When doing calculations by hand, it is tedious to work with such numbers; in a computer this representation would waste valuable storage for insignificant zeros. It is far more convenient to represent these numbers is *scientific notation* or *floating-point notation:*

$$0.598 \times 10^{28}$$

and

$$0.91091 \times 10^{-27}.$$

Each number is thus represented by an ordered pair (f, e), where

$$f = 0.598, \qquad e = 28$$

and

$$f = 0.91091, \qquad e = -27,$$

respectively. The advantage of this notation is obvious: the pair (f, e) can be stored more compactly than a number with many leading or trailing zeros. A number represented by a pair (f, e), as above, is called a *floating-point number*. It is called *normalized* if

$$0.1 \leqq |f| < 1,$$

that is, when the first nonzero digit of the fractional part f immediately follows the decimal point. Of course, there is no need to be restricted to base 10; in fact, most computers use base 2, so that the pair (f, e) represents $f \times 2^e$ and the number is normalized when

$$\frac{1}{2} \leqq |f| < 1.$$

For convenience, our discussion will use base 10.

The most common method of storing a floating-point number in a computer word is to allocate a small number of digits at the left-hand end of the word to represent the exponent; the remaining digits are used for the fractional part. For example, *eeffff* would be $.ffff \times 10^{ee}$. Clearly, the size of the exponent is limited, and the violation of these limits is known as a *floating-point overflow* (exponent too big) or as a *floating-point underflow* (exponent too small). For purposes of example, we shall assume that two decimal digits and a sign are allocated for the exponent while four decimal digits and a sign are allocated for the fractional part. With this assumption

$$-99 \leqq e \leqq 99.$$

Since this representation of real numbers is only approximate, we must carefully examine the usual arithmetic operations to see what effect the approximation has on them. It turns out that most of the familiar laws obeyed by real numbers are *not* obeyed by floating-point numbers; for example, addition of floating-point numbers is not associative, that is, $a + (b + c)$ is not always equal to $(a + b) + c$. The reason for such differences is the rounding of real numbers so they fit in a finite number of digits. This *roundoff error* is the bane of computer programmers! To explain the problem, suppose that we want to add 0.1481×10^2 to 0.5481×10^0. We shall assume that addition is performed in a double-length accumulator as follows: the smaller number, 0.5481×10^0, is placed into the accumulator,

$$0.54810000 \times 10^0,$$

and shifted to the right until the exponent is the same as that in the other number,

$$0.00548100 \times 10^2.$$

The numbers are now added

$$0.00548100 \times 10^2$$
$$0.14810000 \times 10^2$$
$$\overline{0.15358100 \times 10^2}$$

and the result is rounded to four digits:

$$0.1536 \times 10^2.$$

Clearly, this operation is similar, but not identical, to the usual operation of addition. It is an unhappy fact that not only is this operation nonassociative, but it also does not satisfy such statements as "$x + y = x$ if and only if $y = 0$," or "$x + y = 0$ implies $x = -y$." To demonstrate nonassociativity, consider the addition of

$$a = 0.4561 \times 10^{-3},$$
$$b = 0.2524 \times 10^{-3},$$
$$c = 0.1333 \times 10^0.$$

If we add $(a + b) + c$, we get

$$0.1340 \times 10^0,$$

while if we add $a + (b + c)$, we get

$$0.1341 \times 10^0.$$

Suppose that we add 0.1156×10^2 and 0.6134×10^{-3}. Shifting the smaller number to the right yields

$$0.00000613 \times 10^2$$

and adding we get

$$0.11560613 \times 10^2.$$

When this result is rounded to four places it becomes 0.1156×10^2, and so we have an instance of $x + y = x$ but $y \neq 0$. If we add 0.5692×10^{-98} and -0.5623×10^{-98}, we obtain

$$0.00690000 \times 10^{-98};$$

in trying to normalize this we get an exponent less than -99—an underflow. In many computers the result is considered to be zero, and this shows that we cannot conclude that $x = -y$ from $x + y = 0$.

What is the exact relationship between exact addition of real numbers and floating-point addition? If floating-point addition is done in a double-length accumulator, the computed sum of two floating-point numbers is the same as that obtained by calculating the exact sum, normalizing, and rounding to four places. Thus it can be shown that

$$\text{computed value of } a + b = (\text{exact value of } a + b) \cdot (1 + \epsilon),$$

where

$$|\epsilon| \leq 0.5 \times 10^{-3}.$$

In general, for t place decimal floating-point numbers this inequality is

$$|\epsilon| \leq 0.5 \times 10^{1-t}.$$

The value $|\epsilon|$ is called the *relative error due to roundoff* or, more simply, the *relative roundoff error*. There are analogous results for the operation of floating-point subtraction; they will be left to the exercises.

Again assuming that a double-length accumulator is available, multiplication is performed as in the following example: to multiply 0.7583×10^{-4} and -0.5144×10^7, the exponents are added and the fractional parts are multiplied to give the eight-digit product

$$-0.39006952 \times 10^3.$$

This is normalized (if necessary) and rounded to

$$-0.3901 \times 10^3.$$

As with addition, this computed value is the same as the exact value rounded

to four places, and so

computed value of $a \cdot b = $ (exact value of $a \cdot b) \cdot (1 + \epsilon)$,

where

$$|\epsilon| \leq 0.5 \times 10^{-3}$$

or, for t digit arithmetic,

$$|\epsilon| \leq 0.5 \times 10^{1-t}.$$

Finally, let us consider floating-point division. The quotient $0.8315 \times 10^{-5}/0.2456 \times 10^{-3}$ is computed by adjusting the exponents together so that the exponent of the denominator is zero:

$$\frac{0.83150000 \times 10^{-2}}{0.24560000 \times 10^{0}}.$$

Then the numerator is shifted (if necessary) so that its fractional part is smaller than the fractional part of the denominator:

$$\frac{0.08315000 \times 10^{-1}}{0.24560000 \times 10^{0}}.$$

Finally, the fractional parts are divided to give

$$0.33855863 \times 10^{-1}$$

and this is rounded to four places, yielding

$$0.3386 \times 10^{-1}.$$

Again with division, the computed result is the exact result rounded to four places and

computed value of $\dfrac{a}{b} = \left($ exact value of $\dfrac{a}{b} \right) \cdot (1 + \epsilon)$,

where

$$|\epsilon| \leq 0.5 \times 10^{-3}$$

or, for t digit arithmetic,

$$|\epsilon| \leq 0.5 \times 10^{1-t}.$$

5.1.2. Convergence: Fast, Slow, or Never?

Analysis is frequently concerned with determining whether or not a sequence

$$x_1, x_2, x_3, \ldots$$

converges, and, if so, to what limit. In numerical computation we are not only concerned with whether or not a series converges, but how quickly it converges. This section discusses that question.

Consider the two series

$$e^x = 1 + x + \frac{x^2}{2!} + \frac{x^3}{3!} + \cdots$$

and

$$\ln(1 + x) = x - \frac{x^2}{2} + \frac{x^3}{3} - \frac{x^4}{4} + \cdots, \qquad \text{for } -1 < x \leq 1.$$

The infinite sums of these series are the limits of sequences

$$s_1(k) = 1 + x + \cdots + \frac{x^k}{k!},$$

$$s_2(k) = x - \frac{x^2}{2} + \cdots + \frac{(-1)^{k+1}x^k}{k}.$$

Each of these sequences can be easily computed by an iterative procedure. For example, the series for e^x is computed by

Step 1 (Initialize)
> Set $s_1(1) \leftarrow t_1(1) \leftarrow 1$, and $k \leftarrow 0$. We use $t_1(k)$ to denote the kth term in the series.

Step 2 (Next term)
> Compute $k \leftarrow k + 1, t_1(k + 1) \leftarrow t_1(k)x/k$, and $s_1(k + 1) \leftarrow s_1(k) + t_1(k + 1)$; repeat step 2.

The computation of each new partial sum from the previous values uses one multiplication, one division, and two additions. The other sequence can be similarly computed, using the same number of operations (How?).

Since the two sequences can be computed using the same amount of arithmetic per step, the only criterion for deciding which sequence is easier to compute is how many steps must be computed to achieve a certain accuracy. Let

$$E_1(k) = e^x - s_1(k)$$

and

$$E_2(k) = \ln(1 + x) - s_2(k)$$

be the errors after computing k terms of the two series, respectively. The question of how fast the two series for e^x and $\ln(1 + x)$ converge now becomes one of how fast the two sequences $E_1(k)$ and $E_2(k)$ go to zero.

Table 5.1 exhibits the behavior of $s_1(k)$, $s_2(k)$, $E_1(k)$, $E_2(k)$, $E_1(k)/E_1(k-1)$, and $E_2(k)/E_2(k-1)$ for $x = 0.9$; the values were computed with 14-place decimal accuracy although only 10-place accuracy is shown. Even a cursory examination of this table shows a great difference in the rates at which these partial sums converge. Notice that after only 14 steps the series for $e^{0.9}$ has achieved the full 10 digits of accuracy, while even the fiftieth partial sum in the $\ln(1 + 0.9)$ series still differs from the exact value in the fourth decimal place. Clearly, then, the exponential series converges much more quickly than the logarithmic series for $x = 0.9$.

What we would like is a measure of *how quickly* a sequence converges; that is, how much smaller is the $(k+1)$st error than the kth error? A useful such measure is the *order of the convergence*: given a sequence

$$x_1, x_2, x_3, \ldots$$

that converges to a limit α, then *if* there exists a real number $p \geq 1$ such that

$$\lim_{n \to \infty} \frac{|x_{n+1} - \alpha|}{|x_n - \alpha|^p} = c \neq 0,$$

we say that the convergence is of *order p*. The constant C is called the *asymptotic error constant*. When $p = 1$, the series is said to converge *linearly*; in this case the fact that the series converges implies that $C < 1$. For $p = 2$, the convergence is called *quadratic*. The requirement that the constant C be nonzero means that p is as large as possible; without the restriction $C \neq 0$, a quadratically convergent sequence would also satisfy the definition of a linearly convergent sequence (Why?). Notice that if $p > 1$ then C need not be less than 1 for the sequence to converge.

Using some elementary calculus, we can calculate the values of p and C for both the exponential and logarithmic sequences. The major tool we need is a special form of Taylor's theorem, which can be found in any introductory calculus text:

Theorem

Let f be a function that is continuous along with its first $n + 1$ derivatives on the interval $[0, x]$ ($[x, 0]$ if $x < 0$). Then

$$f(x) = f(0) + xf'(0) + \frac{x^2}{2!}f''(0) + \frac{x^3}{3!}f'''(0) + \cdots + \frac{x^n}{n!}f^{(n)}(0) + R_n(x),$$

where $R_n(x) = \int_0^x \frac{(x-t)^n}{n!}f^{(n+1)}(t)\,dt = \frac{x^{n+1}}{(n+1)!}f^{(n+1)}(\theta x)$

for some θ, $0 < \theta < 1$.

Table 5.1 THE CONVERGENCE OF TWO POWER SERIES EXPANSIONS.

Partial sums of the series $e^{0.9} \approx 2.459603111157^-$

k	$s_1(k)$	$E_1(k)$	$E_1(k)/E_1(k-1)$
1	1.0000000000E 00	1.4596031112E 00	
2	1.9000000000E 00	5.5960311116E−01	3.8339402464E−01
3	2.3050000000E 00	1.5460311116E−01	2.7627278704E−01
4	2.4265000000E 00	3.3103111157E−02	2.1411672061E−01
5	2.4538375000E 00	5.7656111570E−03	1.7417127743E−01
6	2.4587582500E 00	8.4486115695E−04	1.4653453623E−01
7	2.4594963625E 00	1.0674865695E−04	1.2635053236E−01
8	2.4595912627E 00	1.1848478379E−05	1.1099416815E−01
9	2.4596019389E 00	1.1722082902E−06	9.8933234521E−02
10	2.4596030066E 00	1.0458128141E−07	8.9217319384E−02
11	2.4596031027E 00	8.4948508228E−09	8.1227258913E−02
12	2.4596031105E 00	6.3323390975E−10	7.4543264263E−02
13	2.4596031111E 00	4.3612669032E−11	6.8872920986E−02
14	2.4596031112E 00	2.7928770407E−12	6.4038205016E−02
15	2.4596031112E 00	1.6875389974E−13	6.0422960725E−02
16	2.4596031112E 00	1.1324274851E−14	6.7105263158E−02
17	2.4596031112E 00	2.6645352591E−15	2.3529411765E−01
18	2.4596031112E 00	2.2204460493E−15	8.3333333333E−01
19	2.4596031112E 00	2.2204460493E−15	1.0000000000E 00
20	2.4596031112E 00	2.2204460493E−15	1.0000000000E 00
21	2.4596031112E 00	2.2204460493E−15	1.0000000000E 00
22	2.4596031112E 00	2.2204460493E−15	1.0000000000E 00
23	2.4596031112E 00	2.2204460493E−15	1.0000000000E 00
24	2.4596031112E 00	2.2204460493E−15	1.0000000000E 00
25	2.4596031112E 00	2.2204460493E−15	1.0000000000E 00
26	2.4596031112E 00	2.2204460493E−15	1.0000000000E 00
27	2.4596031112E 00	2.2204460493E−15	1.0000000000E 00
28	2.4596031112E 00	2.2204460493E−15	1.0000000000E 00
29	2.4596031112E 00	2.2204460493E−15	1.0000000000E 00
30	2.4596031112E 00	2.2204460493E−15	1.0000000000E 00
31	2.4596031112E 00	2.2204460493E−15	1.0000000000E 00
32	2.4596031112E 00	2.2204460493E−15	1.0000000000E 00
33	2.4596031112E 00	2.2204460493E−15	1.0000000000E 00
34	2.4596031112E 00	2.2204460493E−15	1.0000000000E 00
35	2.4596031112E 00	2.2204460493E−15	1.0000000000E 00
36	2.4596031112E 00	2.2204460493E−15	1.0000000000E 00
37	2.4596031112E 00	2.2204460493E−15	1.0000000000E 00
38	2.4596031112E 00	2.2204460493E−15	1.0000000000E 00
39	2.4596031112E 00	2.2204460493E−15	1.0000000000E 00
40	2.4596031112E 00	2.2204460493E−15	1.0000000000E 00
41	2.4596031112E 00	2.2204460493E−15	1.0000000000E 00
42	2.4596031112E 00	2.2204460493E−15	1.0000000000E 00
43	2.4596031112E 00	2.2204460493E−15	1.0000000000E 00
44	2.4596031112E 00	2.2204460493E−15	1.0000000000E 00
45	2.4596031112E 00	2.2204460493E−15	1.0000000000E 00
46	2.4596031112E 00	2.2204460493E−15	1.0000000000E 00
47	2.4596031112E 00	2.2204460493E−15	1.0000000000E 00
48	2.4596031112E 00	2.2204460493E−15	1.0000000000E 00
49	2.4596031112E 00	2.2204460493E−15	1.0000000000E 00
50	2.4596031112E 00	2.2204460493E−15	1.0000000000E 00

Table 5.1 THE CONVERGENCE OF TWO POWER SERIES EXPANSIONS. (CONT.)

Partial sums of the series for ln $(1 + 0.9) \approx 0.641853886172^+$

$s_2(k)$	$E_2(k)$	$E_2(k)/E_2(k-1)$
9.0000000000E−01	−2.5814611383E−01	
4.9500000000E−01	1.4685388617E−01	−5.6887893447E−01
7.3800000000E−01	−9.6146113828E−02	−6.5470595524E−01
5.7397500000E−01	6.7878886172E−02	−7.0599718980E−01
6.9207300000E−01	−5.0219113828E−02	−7.3983408773E−01
6.0349950000E−01	3.8354386172E−02	−7.6374080005E−01
6.7182762857E−01	−2.9973742399E−02	−7.8149451446E−01
6.1801922732E−01	2.3834658851E−02	−7.9518461638E−01
6.6106594832E−01	−1.9212062149E−02	−8.0605568006E−01
6.2619810431E−01	1.5655781861E−02	−8.1489335916E−01
6.5472634032E−01	−1.2872454147E−02	−8.2221726526E−01
6.3119054561E−01	1.0663340560E−02	−8.2838442752E−01
6.5074335968E−01	−8.8894735045E−03	−8.3364809132E−01
6.3440279364E−01	7.4510925347E−03	−8.3819278284E−01
6.4812886911E−01	−6.2749829383E−03	−8.4215608772E−01
6.3654749293E−01	5.3063932421E−03	−8.4564265661E−01
6.4635759981E−01	−4.5037136401E−03	−8.4873348707E−01
6.3801900896E−01	3.8348772097E−03	−8.5149223867E−01
6.4512875485E−01	−3.2748686727E−03	−8.5396963022E−01
6.3904992212E−01	2.8039640568E−03	−8.5620656490E−01
6.4426035017E−01	−2.4064639971E−03	−8.5823639261E−01
6.3978411880E−01	2.0697673765E−03	−8.6008657473E−01
6.4363757015E−01	−1.7836839799E−03	−8.6177992763E−01
6.4031396836E−01	1.5399178150E−03	−8.6333556410E−01
6.4318556031E−01	−1.3316741358E−03	−8.6476961488E−01
6.4070052881E−01	1.1533573601E−03	−8.6609578804E−01
6.4285422278E−01	−1.0003366030E−03	−8.6732580696E−01
6.4098512409E−01	8.6876208636E−04	−8.6846975682E−01
6.4260930640E−01	−7.5542022303E−04	−8.6953636086E−01
6.4119626779E−01	6.5761838614E−04	−8.7053320270E−01
6.4242697883E−01	−5.7309266056E−04	−8.7146690640E−01
6.4135395264E−01	4.9993353328E−04	−8.7234328354E−01
6.4229041186E−01	−4.3652569043E−04	−8.7316745402E−01
6.4147238719E−01	3.8149898440E−04	−8.7394394594E−01
6.4218757448E−01	−3.3368830274E−04	−8.7467677867E−01
6.4156178560E−01	2.9210057351E−04	−8.7536953231E−01
6.4210977370E−01	−2.5588752353E−04	−8.7602540610E−01
6.4162956307E−01	2.2432309835E−04	−8.7664726772E−01
6.4205067085E−01	−1.9678467776E−04	−8.7723769514E−01
6.4168114878E−01	1.7273739577E−04	−8.7779901230E−01
6.4200560718E−01	−1.5172101026E−04	−8.7833331964E−01
6.4172054730E−01	1.3333887504E−04	−8.7884252034E−01
6.4197113482E−01	−1.1724865203E−04	−8.7932834289E−01
6.4175073170E−01	1.0315446837E−04	−8.7979236078E−01
6.4194468645E−01	−9.0800277586E−05	−8.8023600935E−01
6.4177392195E−01	7.9964227007E−05	−8.8066060075E−01
6.4192434004E−01	−7.0453868528E−05	−8.8106733679E−01
6.4179178409E−01	6.2102078163E−05	−8.8145732037E−01
6.4190864975E−01	−5.4763572797E−05	−8.8183156534E−01
6.4180557424E−01	4.8311931349E−05	−8.8219100548E−01

Since both of the sequences arose from the series given in the theorem, we know that

$$E_1(k) = e^x - s_1(k) = E_1(k-1) - \frac{x^k}{k!}$$

and

$$E_2(k) = \ln(1+x) - s_2(k) = E_2(k-1) + \frac{(-x)^k}{k}.$$

These yield, respectively, that

$$\lim_{k \to \infty} \frac{E_1(k)}{E_1(k-1)} = 1 - \lim_{k \to \infty} e^{-\theta x}, \qquad 0 < \theta < 1,$$

and

$$\lim_{k \to \infty} \frac{E_2(k)}{E_2(k-1)} = 1 - \lim_{k \to \infty} (1+x).$$

Thus the convergence of each sequence is linear ($p = 1$).

What explains the different rates of convergence we observed in Table 5.1? It must be the asymptotic error constants. For the exponential series we have

$$C = 1 - \lim_{k \to \infty} e^{-\theta x} = 1 - \lim_{k \to \infty} e^{-0.9\theta}$$

$$= 1 - e^{-0.9\theta}, \qquad 0 < \theta < 1,$$

so that taking $\theta = 1$ yields

$$C < 1 - e^{-0.9} \approx 0.59343,$$

while for the logarithmic series

$$C = 1 - \lim_{k \to \infty} (1+x) = 1 - (1+x) = -x = -0.9.$$

This explains the difference in the convergence of the series. Notice in Table 5.1 that the value of $E_1(k)/E_1(k-1)$ did not have time to converge to a value, since the series of e^x converged very quickly to the limits of the computer's accuracy, while $E_2(k)/E_2(k-1)$ is clearly displaying its convergence to -0.9, as our analysis indicates that it should.

The notion of the order of convergence is applicable to any iterative process. Suppose that we want to find a root of the equation

$$f(x) = 0$$

for some function $f(x)$. If $f(x)$ is simple enough, such as a quadratic or cubic function of x, there are well-known methods for finding solutions; but if $f(x)$ involves, say, exponential or trigonometric functions, an analytical

solution may not be possible. In the latter case we must rely on computational methods that approximate a root to any degree of accuracy.

Consider the well-known Newton's method of solving $f(x) = 0$. We
make an initial guess, x_0, and then we iterate to improve the accuracy of our
guess: the new approximation is calculated from x_0 by finding the place
where the tangent to $f(x)$ at the point $(x_0, f(x_0))$ crosses the x-axis, as shown
in Figure 5.1(a). From analytical geometry we know that the equation of

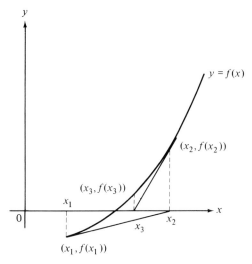

(a) Succesive approximation to the
    root of $f(x) = 0$ by Newton's method

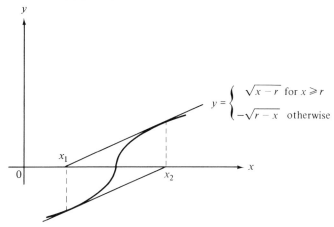

(b) A case in which Newton's
    method does not converge

Fig. 5.1 Newton's method of solving $f(x) = 0$.

that tangent line is

$$y - f(x_0) = f'(x_0)(x - x_0).$$

Now, assuming that $f'(x_0) \neq 0$ (if it is zero, we must pick a different value of x_0), this line crosses the x-axis at the point

$$x_1 = x_0 - \frac{f(x_0)}{f'(x_0)}.$$

In general, we iterate with this method, giving

$$x_{i+1} = x_i - \frac{f(x_i)}{f'(x_i)}.$$

For example, suppose that we wish to solve $f(x) = x^2 - 2 = 0$. We might start out with $x_0 = 1$, and the iteration would proceed to

$$x_1 = x_0 - \frac{f(x_0)}{f'(x_0)} = 1 - \frac{1^2 - 2}{2 \cdot 1} = 1.5,$$

$$x_2 = x_1 - \frac{f(x_1)}{f'(x_1)} = 1.5 - \frac{(1.5)^2 - 2}{2 \cdot 1.5} \approx 1.4166,$$

and so on.

How quickly does it converge? First, Newton's method does *not* always converge. Figure 5.1(b) shows a case where successive approximations repeat. For a function like

$$f(x) = \begin{cases} \sqrt{x - r} & \text{for } x \geq r, \\ -\sqrt{r - x} & \text{otherwise,} \end{cases}$$

if we are unlucky enough to choose the x_1 shown in Figure 5.1(b) as an initial guess, no amount of iteration will improve our estimate of the root of the equation $f(x) = 0$, since $x_i = x_{i+2}$ for all i.

Suppose that $f(x)$ and its first three derivatives are continuous and bounded in the vicinity of the root, α, of $f(x) = 0$. We can estimate the accuracy of Newton's method using a more general form of Taylor's theorem to expand the function $g(x) = x - [f(x)/f'(x)]$ in a Taylor series about the root $x = \alpha$. We get

$$g(x) = g(\alpha) + g'(\alpha)(x - \alpha) + g''(\theta x)\frac{(x - \alpha)^2}{2}, \qquad 0 < \theta < 1.$$

Applying this formula to $x = x_i$, we obtain

$$x_{i+1} = \alpha + \frac{1}{2}g''(\theta x_i)(x_i - \alpha)^2,$$

SO

$$|x_{i+1} - \alpha| = |x_i - \alpha|^2 \cdot \left| \frac{g''(\theta x_i)}{2} \right|.$$

Now, using the boundedness of the derivatives of $f(x)$, it can be shown that $g''(x)$ is bounded on a suitably restricted neighborhood of the root. Thus, when it does converge, Newton's method converges quadratically.

Rather than let the reader think that the order of convergence is always an integer, we mention the fact that one method for solving nonlinear equations (the secant method) has order of convergence $(1 + \sqrt{5})/2 \approx 1.618$.

5.1.3. Stability

Errors, regardless of their original cause, propagate in different ways. Some errors decay and are thus little problem, but others can grow so large as to dominate the computation. The *stability* of a numerical method depends on the growth rate of such errors. Do small errors in the inputs cause only small errors in the outputs? If so, then the method is considered stable; however, if small errors in the inputs have a disastrous effect on the outputs, the method is unstable.

To examine stability in numerical calculation, we shall develop an extended example. Consider the Fibonacci numbers $F_0, F_1, F_2, \ldots$ defined by

$$F_0 = 0,$$
$$F_1 = 1,$$
$$F_{k+2} = F_{k+1} + F_k, \qquad k \geq 0.$$

This last equation is called a *linear recurrence relation*, since the next element in the sequence is given by a linear combination of previous elements. Since such recurrence relations often provide efficient ways of computing sequences, it is important to study their computational properties, especially their stability.

The asymptotic behavior of such a sequence can be determined by the following technique, which we shall apply to the Fibonacci numbers: What is a solution to

$$F_{k+2} = F_{k+1} + F_k$$

in the form

$$F_k = cr^k?$$

Substituting this into the original equation, we get

$$c(r^{k+2} - r^{k+1} - r^k) = 0,$$

so that either $c = 0$ or $r = 0$ (and hence $F_i = 0$ for $i \geq 0$), or r is a solution to the quadratic equation

$$r^2 - r - 1 = 0$$

and c is arbitrary. Thus

$$r = \frac{1 \pm \sqrt{5}}{2} = \begin{cases} 1.618\ldots \\ -0.618\ldots \end{cases}$$

(the positive value is known as the *golden ratio*). Thus for arbitrary c_1 and c_2, both

$$F_k = c_1 \left(\frac{1 + \sqrt{5}}{2}\right)^k$$

and

$$F_k = c_2 \left(\frac{1 - \sqrt{5}}{2}\right)^k$$

are solutions. Since the original equation is linear and homogeneous, the sum or superposition of these two solutions is also a solution:

$$F_k = c_1 \left(\frac{1 + \sqrt{5}}{2}\right)^k + c_2 \left(\frac{1 - \sqrt{5}}{2}\right)^k.$$

Moreover, we assert without proof that *every* solution has this form. The values of F_0 and F_1 give us "initial conditions"

$$F_0 = c_1 + c_2 = 0,$$

$$F_1 = c_1 \left(\frac{1 + \sqrt{5}}{2}\right) + c_2 \left(\frac{1 - \sqrt{5}}{2}\right) = 1,$$

and so

$$c_1 = \frac{1}{\sqrt{5}} \quad \text{and} \quad c_2 = -\frac{1}{\sqrt{5}}.$$

Thus the Fibonacci numbers are given in closed form by *Binet's formula*

$$F_k = \frac{1}{\sqrt{5}} \left(\left(\frac{1 + \sqrt{5}}{2}\right)^k - \left(\frac{1 - \sqrt{5}}{2}\right)^k\right).$$

From this formula we can easily determine the asymptotic behavior of F_k. Notice that $|(1 - \sqrt{5})/2| < 1$, while $(1 + \sqrt{5})/2 > 1$. Thus, as $k \to \infty$, $((1 - \sqrt{5})/2)^k \to 0$, while $((1 + \sqrt{5})/2)^k$ grows exponentially. We conclude that the Fibonacci numbers F_k behave asymptotically like $((1 + \sqrt{5})/2)^k/\sqrt{5}$.

Now let us consider two Fibonacci-type sequences that satisfy the same recurrence relation, but different initial conditions: a *dominant* sequence D_k and a *weak* sequence W_k with initial conditions

$$D_0 = 1 \qquad \text{and} \qquad W_0 = 1$$
$$D_1 = \frac{1 + \sqrt{5}}{2} \qquad \qquad W_1 = \frac{1 - \sqrt{5}}{2}.$$

These sequences are given in closed form by

$$D_k = \left(\frac{1 + \sqrt{5}}{2}\right)^k \quad \text{and} \quad W_k = \left(\frac{1 - \sqrt{5}}{2}\right)^k.$$

Table 5.2 lists the results of an actual computation of these two sequences from the recurrence relations, along with the values of $((1 \pm \sqrt{5})/2)^k$ and the ratio W_k/W_{k-1}, which should be approximately -0.618033 for all k.

An examination of this table indicates that the computation did not progress as expected; in fact, far from it! What happened? Among all the sequences that satisfy the relation

$$S_{k+2} = S_{k+1} + S_k$$

and are hence represented by

$$S_k = c_1\left(\frac{1 + \sqrt{5}}{2}\right)^k + c_2\left(\frac{1 - \sqrt{5}}{2}\right)^k$$

for some constants c_1 and c_2, the sequences in which $c_1 = 0$ are exceptional: they are decreasing in absolute magnitude since $|(1 - \sqrt{5})/2| < 1$; whenever $c_1 \neq 0$, the sequences are increasing in absolute magnitude, since $(1 + \sqrt{5})/2 > 1$. Theoretically, the initial conditions for the sequence W_k demand that $c_1 = 0$ and $c_2 = 1$; however, since we only *approximated* the initial condition $W_1 = (1 - \sqrt{5})/2$ by $W_1 = -0.618033$, the values of c_1 and c_2 we are *actually* using are not exactly 0 and 1, but rather $c_1 = \epsilon$ and $c_2 = 1 + \delta$ for very small ϵ and δ. Thus we are actually computing

$$W_k' = \epsilon\left(\frac{1 + \sqrt{5}}{2}\right)^k + (1 + \delta)\left(\frac{1 - \sqrt{5}}{2}\right)^k,$$

and no matter how small ϵ is, the sequence will behave asymptotically like $((1 + \sqrt{5})/2)^k$ instead of like $((1 - \sqrt{5})/2)^k$. This is precisely the behavior exhibited in Table 5.2.

The computational difficulty we have encountered is a typical example of instability: a small error grows disproportionately in comparison to the correct solution. Our calculations were done using floating-point numbers

Table 5.2 AN EXAMPLE OF INSTABILITY.

k	$\left(\dfrac{1+\sqrt{5}}{2}\right)^k$	D_k	W_k	$\dfrac{W_k}{W_{k-1}}$	$\left(\dfrac{1-\sqrt{5}}{2}\right)^k$
0	1.00000E+00	1.00000E+00	1.00000E+00		1.00000E+00
1	1.61803E+00	1.61803E+00	−6.18033E−01	−6.18033E−01	−6.18033E−01
2	2.61803E+00	2.61803E+00	3.81966E−01	−6.18034E−01	3.81965E−01
3	4.23606E+00	4.23606E+00	−2.36067E−01	−6.18033E−01	−2.36067E−01
4	6.85408E+00	6.85410E+00	1.45898E−01	−6.18035E−01	1.45897E−01
5	1.10901E+01	1.10901E+01	−9.01694E−02	−6.18029E−01	−9.01698E−02
6	1.79442E+01	1.79442E+01	5 57289E−02	−6.18046E−01	5.57280E−02
7	2.90343E+01	2.90344E+01	−3.44405E−02	−6.18000E−01	−3.44418E−02
8	4.69785E+01	4.69786E+01	2.12883E−02	−6.18120E−01	2.12862E−02
9	7.60128E+01	7.60131E+01	−1.31521E−02	−6.17807E−01	−1.31555E−02
10	1.22991E+02	1.22991E+02	8.13627E−03	−6.18628E−01	8.13059E−03
11	1.99003E+02	1.99004E+02	−5.01585E−03	−6.16480E−01	−5.02498E−03
12	3.21994E+02	3.21996E+02	3.12042E−03	−6.22112E−01	3.10561E−03
13	5.20998E+02	5.21001E+02	−1.89542E−03	−6.07426E−01	−1.91937E−03
14	8.42992E+02	8.42998E+02	1.22499E−03	−6.46289E−01	1.18623E−03
15	1.36398E+03	1.36399E+03	−6.70433E−04	−5.47294E−01	−7.33135E−04
16	2.20698E+03	2.20699E+03	5.54561E−04	−8.27169E−01	4.53102E−04
17	3.57097E+03	3.57099E+03	−1.15871E−04	−2.08942E−01	−2.80032E−04
18	5.77794E+03	5.77799E+03	4.38690E−04	−3.78600E+00	1.73069E−04
19	9.34890E+03	9.34898E+03	3.22818E−04	7.35869E−01	−1.06962E−04
20	1.51268E+04	1.51269E+04	7.61508E−04	2.35893E+00	6.61006E−05
21	2.44757E+04	2.44759E+04	1.08432E−03	1.42391E+00	−4.08561E−05
22	3.96025E+04	3.96029E+04	1.84583E−03	1.70228E+00	2.52504E−05
23	6.40783E+04	6.40789E+04	2.93016E−03	1.58744E+00	−1.56056E−05
24	1.03680E+05	1.03681E+05	4.77600E−03	1.62994E+00	9.64482E−06
25	1.67758E+05	1.67760E+05	7.70616E−03	1.61351E+00	−5.96082E−06
26	2.71438E+05	2.71442E+05	1.24821E−02	1.61976E+00	3.68399E−06
27	4.39197E+05	4.39203E+05	2.01883E−02	1.61737E+00	−2.27683E−06
28	7.10636E+05	7.10645E+05	3.26704E−02	1.61828E+00	1.40715E−06
29	1.14983E+06	1.14984E+06	5.28588E−02	1.61793E+00	−8.69672E−07
30	1.86046E+06	1.86049E+06	8.55293E−02	1.61807E+00	5.37486E−07
31	3.01029E+06	3.01034E+06	1.38388E−01	1.61801E+00	−3.32185E−07
32	4.87077E+06	4.87083E+06	2.23917E−01	1.61803E+00	2.05301E−07
33	7.88106E+06	7.88117E+06	3.62305E−01	1.61803E+00	−1.26883E−07
34	1.27518E+07	1.27520E+07	5.86223E−01	1.61803E+00	7.84182E−08
35	2.06328E+07	2.06331E+07	9.48528E−01	1.61803E+00	−4.84651E−08
36	3.33846E+07	3.33851E+07	1.53475E−00	1.61803E+00	2.99531E−08
37	5.40174E+07	5.40183E+07	2.48328E−00	1.61803E+00	−1.85120E−08
38	8.74020E+07	8.74035E+07	4.01803E−00	1.61803E+00	1.14410E−08
39	1.41419E+08	1.41421E+08	6.50131E−00	1.61803E+00	−7.07096E−09
40	2.28821E+08	2.28825E+08	1.05193E−01	1.61803E+00	4.37008E−09

with 23 bits in the fractional part; higher precision would have delayed the growth in the error, but it could not have prevented it. Thus, while the Fibonacci recurrence relation can be used to compute the sequence D_k, it is practically useless for the computation of W_k. To compute W_k, we have to resort to some stable computational method, such as the use of the closed form $W_k = ((1 - \sqrt{5})/2)^k$.

5.2. COMPUTATION OF MATHEMATICAL CONSTANTS

In this section we discuss the difficulty of computing the values of various mathematical constants (like π or e) to ultrahigh precision. We are interested not in finding the first 10, 50, or even 100 decimal digits in the fractional parts of these numbers; we want to compute the first several hundred thousand or even one million digits of these numbers. The combinatorial problems involved in such computations are difficult to deal with; one must begin by writing subroutines to perform arithmetic operations on very large (or very precise) numbers. Since most modern computers are word oriented, such high-precision arithmetic routines can be difficult enough by themselves. Once the subroutines are written and debugged, algorithms must be designed to actually do the calculation. Since each stage of the calculation will involve several (perhaps many) high-precision arithmetic operations, each of which involves a fairly lengthy procedure, the rate of convergence of the method is critical, as is the amount of work to be performed at each stage.

Having solved all these problems and then having computed some constant to, say, one million decimal places, how do we know that it is correct? For a number like $\sqrt{2}$, we need only square the result and compare it to 2, but what if we are trying to check a computation of π or e to one million decimal places? These constants have no rational (or even algebraic) relation to any rational (algebraic) numbers, so no simple check of the final result is possible—the check must be incorporated into every stage of the computation. Determining a way of checking the results of a computation can be as difficult a problem as the computation itself.

Why go to all the trouble of calculating π, for example, to one million decimal places? Certainly, the first 100 decimal digits, known for centuries, are sufficient for the solution to any real-life problems such as are found in physics or astronomy. Why should we expend great effort on unneeded precision? Initially, during the time of the Greeks perhaps, the reason for such precise determinations of π and $\sqrt{2}$ was the hope that they would be found to be *rational*. Even after it was proved that these numbers do not have finite or repeating decimal expansions, the calculations continued—it had become a grand sport. More recently, additional motivation has been found: we want to determine whether or not the decimal digits of these numbers are random or not; that is, is the sequence of digits in π ($e, \sqrt{2}$, etc.) a perfect infinite random sequence, as discussed in Section 4.1.1? By studying the statistical properties of very many digits of a number, it is hoped that patterns can be found and conjectures formulated and proved. Unfortunately, in spite of all the statistical analysis that has been performed, no conjectures have yet been proved about the distribution of digits in "naturally occurring" irrational numbers.

After centuries of investigation, there are easily stated questions about the nature of familiar numbers to which no one knows the answer. For example, no one knows whether every digit occurs infinitely often in the decimal expansion of π. Of course, it is conjectured that this is the case, but there is no known proof. As long as we have no firm answers to such questions, it is not surprising that people resort to gathering experimental evidence in the hope of finding some clue to the answer.

5.2.1. $\sqrt{2}$

The earliest high-precision calculation of $\sqrt{2}$ appears to have been done (by hand) in the late nineteenth century by J. Marcus Boorman, a "Consultative Mechanician, and Attorney and Counselor at Law." His method, which is difficult to decipher, is close to the usual high school method for calculating square roots. Boorman calculated $\sqrt{2}$ to 568 decimal places, claiming to have verified the first 486. A few months later Artemas Martin described a method for improving an estimate r to $\sqrt{N}$: let $R = N - r^2$; then

$$\sqrt{N} = \frac{N}{r}\sqrt{1 - \frac{R}{N}},$$

and this square root can be approximated by the binomial series

$$\sqrt{1 - x} = 1 - \left(\frac{x}{2}\right) - \frac{1}{1\cdot 2}\left(\frac{x}{2}\right)^2 - \cdots - \frac{1\cdot 3\cdot 5 \cdots (2n - 3)}{n!}\left(\frac{x}{2}\right)^n - \cdots,$$

and the result is an improved value for $\sqrt{N}$. Martin showed how to apply his method to $\sqrt{2}$ by starting with a 19-digit approximation and then, in one iteration, his method yielded 145 correct digits. Martin did not indicate how many terms of the binomial series he had to use.

Next in the $\sqrt{2}$ competition are René Constal and Horace Uhler who, at first independently, each computed $\sqrt{2}$ to more than 1000 decimal digits in 1950. Constal, who did not indicate by what means he performed the calculations, used the binomial-like series

$$S(x) = 1 + x + \frac{1\cdot 3}{1\cdot 2}x^2 + \frac{1\cdot 3\cdot 5}{1\cdot 2\cdot 3}x^3 + \cdots + \frac{1\cdot 3\cdot 5\ldots(2n - 1)}{n!}x^n + \cdots,$$

and thus we define

$$P_{k+1} = P_k^2 - 2$$

and

$$Q_{k+1} = P_k Q_k.$$

This is again a solution to the Pell equation since

$$\begin{aligned}
P_{k+1}^2 - NQ_{k+1}^2 &= (P_k^2 - 2)^2 - N(P_k Q_k)^2 \\
&= P_k^4 - 2P_k^2 + 4 - NP_k^2 Q_k^2 \\
&= P_k^2(P_k^2 - NQ_k^2 - 4) + 4 \\
&= 4.
\end{aligned}$$

We leave to Exercise 11 the proof that

$$\lim_{k \to \infty} \frac{P_k}{Q_k} = \sqrt{N} \cdot$$

Dutka chose $P_0 = 6726$ and $Q_0 = 4756$ and then performed 17 iterations of his method. The integers P_{17} and Q_{17} each contain over half a million digits, and their quotient is $\sqrt{2}$ correct to over one million digits. Dutka then checked his result by squaring it; the square was a 1 followed by a decimal point and 1,000,082 nines.

5.2.2. e

During the seventeenth and eighteenth centuries much attention was paid to the study of the compound-interest law,

$$A = P\left(1 + \frac{r}{n}\right)^{nt},$$

where p is the principal, r is the interest rate, n is the number of times per year the interest is compounded, and A is the value of the annuity after t years. Thus

$$\lim_{n \to \infty} \left(1 + \frac{1}{n}\right)^n$$

can be interpreted as being the amount accrued after one year from the investment of $1.00 at 100 per cent/year, compounded continuously. This same limit also arises naturally in the study of logarithms. We define the function $\log_b x$ to have the value y for which $b^y = x$. Now, the derivative dy/dx of a function

$y = f(x)$ is defined as

$$\lim_{\Delta x \to 0} \frac{f(x + \Delta x) - f(x)}{\Delta x}$$

so that

$$\frac{d}{dx} (\log_b x) = \lim_{\Delta x \to 0} \frac{\log_b (x + \Delta x) - \log_b (x)}{\Delta x}$$

$$= \lim_{\Delta x \to 0} \frac{\log_b ((x + \Delta x)/x)}{\Delta x}$$

$$= \lim_{\Delta x \to 0} \log_b \left(1 + \frac{\Delta x}{x}\right)^{1/\Delta x}$$

$$= \lim_{\Delta x \to 0} \frac{1}{x} \log_b \left(1 + \frac{\Delta x}{x}\right)^{x/\Delta x}$$

$$= \frac{1}{x} \log_b \left(\lim_{t \to 0} (1 + t)\right)^{1/t}.$$

This value, $\lim_{t \to 0} (1 + t)^{1/t}$ was known to Napier, Oughtred, and other seventeenth century mathematicians. Using the binomial theorem, they were able to show that

$$e = \lim_{t \to 0} (1 + t)^{1/t} = \lim_{n \to \infty} \left(1 + \frac{1}{n}\right)^n = 1 + \frac{1}{1!} + \frac{1}{2!} + \frac{1}{3!} + \cdots .$$

The use of the symbol e to represent this limit goes back to Euler in 1727: He calculated e correctly to 23 decimal places, presumably using the factorial series.

Unlike π (see the following section) there seems to have been little interest in computing the value of e. In fact, it appears that after Euler, the first large-scale computation of e was made at the suggestion of John E. von Neumann by G. Reitwiesner in 1949. He computed the value of e to about 2500 decimal places, using the factorial series on the ENIAC computer. In 1952 F. Gruenberger, using Reitwiesner's intermediate output, extended the computation of e to 3000 decimal places, using an IBM 602A calculating card punch. In 1953 D. Wheeler computed 60,000 digits of e on the ILLIAC I. Then, in 1961, Daniel Shanks and John Wrench, using an IBM 7090, computed 100,265 digits. Finally, in 1964 Donald Gillies and D. Wheeler computed over one million digits on the ILLIAC II.

The method used in each of these computations was basically the same: the summation of the terms in the factorial series. Wheeler used the first 16,000 terms, Shanks and Wrench used the first 25,266 terms, and Gillies and Wheeler used almost 200,000 terms. To illustrate some of the techniques needed for the computation, we shall discuss Wheeler's approach in detail.

Write

$$e - 1 = 1 + \frac{1}{2!} + \frac{1}{3!} + \frac{1}{4!} + \cdots$$

$$= 1 + \frac{1}{2}\left(1 + \frac{1}{3}\left(1 + \frac{1}{4}\left(1 + \frac{1}{5}\left(1 + \cdots\right)\right)\right)\right).$$

The computation proceeds from the innermost division to the outermost. Each division is done exactly, and the remainders are saved so as to allow the computation to be extended. For example, to sum the first six terms in the series to three-decimal-place accuracy,

$$e - 1 \approx 1 + \frac{1}{2}\left(1 + \frac{1}{3}\left(1 + \frac{1}{4}\left(1 + \frac{1}{5}\left(1 + \frac{1}{6}\right)\right)\right)\right).$$

Step	Computation
1	$1 + \frac{1}{6}\cdot 1$ $= 1.166$, remainder 4 out of 6
2	$1 + \frac{1}{5}(1.166) = 1.233$, remainder 1 out of 5
3	$1 + \frac{1}{4}(1.233) = 1.308$, remainder 1 out of 4
4	$1 + \frac{1}{3}(1.308) = 1.436$, remainder 0 out of 3
5	$1 + \frac{1}{2}(1.436) = 1.718$, remainder 1 out of 2

Each division operation is checked by a variant of the rule of *casting out nines*; that is, we compute the sum of the digits in the quotient modulo 9 ($1 + 6 + 6 = 13 \equiv 4$ in step 1) and then we compute, in the following step, the quotient times the divisor plus the remainder minus 1, also modulo 9 ($5\cdot 233 + 1 - 1 = 1165 \equiv 4$ in step 2); Exercise 13 is to show that this does indeed check the computation. Now, suppose that we want to extend this computation to the first *five* decimal places; we use exactly the same process, but adding the remainders to the quotient, instead of adding 1:

Step	Remainder	Computation
0	—	$4 + \frac{1}{7}\cdot 0$ $= 4.00$
1	4	$1 + \frac{1}{6}(4.00) = 1.66$, new remainder 4 out of 6
2	1	$1 + \frac{1}{5}(1.66) = 1.33$, new remainder 1 out of 5
3	1	$0 + \frac{1}{4}(1.33) = 0.33$, new remainder 1 out of 4
4	0	$1 + \frac{1}{3}(0.33) = 1.11$, new remainder 0 out of 3
5	1	$0 + \frac{1}{2}(1.11) = 0.55$, new remainder 1 out of 2

Now, from this we conclude that, correct to five decimal places,

$$1 + \frac{1}{2}\left(1 + \frac{1}{3}\left(1 + \frac{1}{4}\left(1 + \frac{1}{5}\left(1 + \frac{1}{6}\right)\right)\right)\right) \approx 1.71855.$$

Of course, this part of the computation can be checked by a method similar to that used on the earlier part.

In actual practice, Wheeler used not decimal arithmetic, but arithmetic base 10^{10}. Hence each calculation, which above yielded only two or three decimal digits, actually produced 10 decimal digits; the check was then done modulo $10^{10} - 1$.

5.2.3. π

Man has been interested in the ratio of the circumference for a circle to its diameter since biblical times. In fact, the approximation $\pi \approx 3$ can be inferred from I Kings 7:23, which states

> And he made a molten sea, ten cubits from the one brim to the other; it was round all about ... and a line of thirty cubits did compass it round about.

The ancient Chinese also used the approximation $\pi \approx 3$, and the ancient Egyptians (in the famed Ahmes Papyrus) assumed that the area of a circle is the same as a square of $\frac{8}{9}$ of the diameter; this implies that $\pi = \frac{256}{81} \approx 3.16$.

It was the ancient Greeks, however, who made the first attempts at computing π more exactly; they were trying to square the circle, which is, of course, equivalent to the exact determination of π by geometric means. Archimedes (ca. 225 B.C.) determined that $3\frac{1}{7} > \pi > 3\frac{10}{71}$ by computing the areas of regular polygons with 96 sides inscribed and circumscribed about the

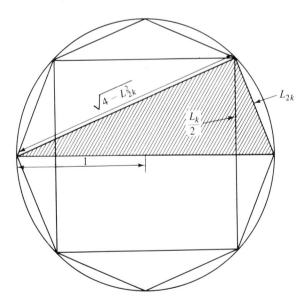

Fig. 5.2 Archimedes' method for approximating π.

unit circle. To understand how Archimedes arrived at the lower bound, consider the diagram shown in Figure 5.2. Here L_k is the length of a side of a regular polygon with k sides, inscribed in the unit circle. Since twice the area of the shaded triangle can be written either as

$$L_{2k}\sqrt{4 - L_{2k}^2}$$

or as

$$2 \cdot \frac{L_k}{2},$$

we can find L_{2k} in terms of L_k by solving the resulting equation; we find that

$$L_{2k} = \sqrt{2 - \sqrt{4 - L_k^2}}.$$

Since the side of a hexagon inscribed in a circle is equal to the radius of the circle, we know that

$$L_6 = 1.$$

Using this and the relation between L_{2k} and L_k, we compute

$$L_{12} = \sqrt{2 - \sqrt{4 - 1}} = \sqrt{2 - \sqrt{3}},$$

$$L_{24} = \sqrt{2 - \sqrt{4 - 2 + \sqrt{3}}} = \sqrt{2 - \sqrt{2 + \sqrt{3}}},$$

$$L_{48} = \sqrt{2 - \sqrt{4 - 2 + \sqrt{2 + \sqrt{3}}}} = \sqrt{2 - \sqrt{2 + \sqrt{2 + \sqrt{3}}}},$$

and finally

$$L_{96} = \sqrt{2 - \sqrt{4 - 2 + \sqrt{2 + \sqrt{2 + \sqrt{3}}}}}$$
$$= \sqrt{2 - \sqrt{2 + \sqrt{2 + \sqrt{2 + \sqrt{3}}}}} \approx 0.06544.$$

We have

$$\pi r^2 = \pi = \text{area of circle} > \text{area of polygon}$$

$$= 48 L_{96} \sqrt{\frac{1 - L_{96}^2}{4}} \approx 3.1394 \approx 3\frac{10}{71}.$$

The derivation of Archimedes' upper bound is left to Exercise 14; this value, $\frac{22}{7}$, is still accepted as a satisfactory approximation to π—it is in error by only 0.2 per cent.

The Archimedean method for determining the value of π remained popular for almost 1800 years; in 1579 François Viète found π to nine decimal places by using polygons with $6 \cdot 2^{16} = 393,216$ sides, and Adrian Van Roomen used 2^{30}-gons to find π to 15 decimal places in 1593. Ludolph Van Ceulen died of exhaustion after using 2^{62}-gons to compute 35 decimal places of π in 1610; his achievement was so extraordinary that the number was engraved on his tombstone!

With Van Ceulen the use of geometric means to compute π came to an end, since the tools of the new subject of calculus yielded far more powerful techniques of approximation, some of which are still used on computers today. John Wallis, in 1655, discovered the infinite product

$$\frac{\pi}{2} = \frac{2}{1} \cdot \frac{2}{3} \cdot \frac{4}{3} \cdot \frac{4}{5} \cdot \frac{6}{5} \cdot \frac{6}{7} \cdot \frac{8}{7} \cdot \frac{8}{9} \cdots,$$

and Lord Brouncker found the infinite continued fraction

$$\frac{\pi}{4} = \cfrac{1}{1 + \cfrac{1^2}{2 + \cfrac{3^2}{2 + \cfrac{5^2}{2 + \cdot}}}}$$

The most important contribution is that of James Gregory in 1671, who discovered the power series

$$\arctan x = x - \frac{x^3}{3} + \frac{x^5}{5} - \cdots, \qquad \text{for } -1 \leq x \leq 1.$$

This series has been used in virtually every major calculation of π for the last three centuries! Using the fact $\arctan(1/\sqrt{3}) = \pi/6$, Abraham Sharp computed π to 72 decimal places in 1699, doubling the then known accuracy; his computation was extended in 1719 to 127 places by Fautet de Lagny (there was an error in the 113th place of his value).

Combining Gregory's series with the identity

$$\arctan x + \arctan y = \arctan \frac{x + y}{1 - xy}$$

yields the technique most used to compute π. For example, in 1794 Baron Georg von Vega used

$$\frac{\pi}{4} = 5 \arctan \frac{1}{7} + 2 \arctan \frac{3}{79}$$

in conjunction with Gregory's series to compute π to 140 decimal places. Similarly, Zaharias Dahse used

$$\frac{\pi}{4} = \arctan \frac{1}{2} + \arctan \frac{1}{5} + \arctan \frac{1}{8}$$

to compute 200 places. Rather than mention every computation made by hand in those precomputer days, we shall mention only one more, the most spectacular one; the reader is urged to see the papers and books cited in Section 5.4 for a more complete history. In 1873, William Shanks published the first 707 decimal places of π, which he had calculated by the identity

$$\frac{\pi}{4} = 4 \arctan \frac{1}{5} - \arctan \frac{1}{239},$$

discovered by John Machin in 1706. Unfortunately, Shanks's value was in error beyond the 527th place.

Using Machin's formula on a desk calculator, Smith and Wrench in 1945–1948 computed 808 places of which the first 722 were correct. The first real computer computation was by G. Reitwiesner in 1949; he used ENIAC, the world's first electronic computer, to find 2037 places. S. C. Nicholson and J. Jeenel computed 3089 places on NORC in 1954–1955. In 1958 F. Genuys computed 10,000 places on an IBM 704 and then in 1959 extended this to 16,167 places. Daniel Shanks (no relation to William) and John Wrench computed 100,000 places on an IBM 7094 in 1961. J. Gilloud and Fillatoire computed 250,000 places on a STRETCH computer in 1966, and then in 1967 Gilloud and M. Dichampt computed 500,000 places on a CDC 6600. Although Shanks and Wrench's calculation has been surpassed, we shall discuss it in detail, since the methods used by Gilloud and his colleagues have never been published.

Shanks and Wrench used the identity

$$\pi = 24 \arctan \frac{1}{8} + 8 \arctan \frac{1}{57} + 4 \arctan \frac{1}{239},$$

which was discovered by Störmer in 1896. Each of the three arctangents was computed using a modified Gregory series; instead of computing

$$A \arctan \frac{1}{m} = \sum_{k=0}^{\infty} \frac{(-1)^k Am}{(2k+1)m^{2(k+1)}}$$

one term at a time (which would require four divisions, one subtraction, and one addition), they computed two terms at a time by using

$$A \arctan \frac{1}{m} = \sum_{k=0}^{\infty} \frac{Am[(4k+3)m^2 - (4k+1)]}{(16k^2 + 16k + 3)m^{4(k+1)}},$$

which requires two divisions, one multiplication, and one addition, a savings of 27 per cent of the computation time based on the speed of operations on the IBM 7090. Since the computer was a binary machine, a further savings was effected in the computation of $24 \arctan \frac{1}{8}$, where multiplication by $((4k + 3)m^2 - (4k + 1))$ could be done by simply shifting the multiplicand.

At any given time, their program required only three numbers: the current value of $Am/m^{4(k+1)}$, the current term in the sum, and the current partial sum. Since an IBM 7090 word holds the equivalent of more than 10 decimal digits, the 32,768 words of memory were sufficient to store each of the three numbers in memory to more than 100,000 decimal places.

They checked their computation using Gauss's identity

$$\pi = 48 \arctan \frac{1}{18} + 32 \arctan \frac{1}{57} - 20 \arctan \frac{1}{239}.$$

Since they had already computed $8 \arctan \frac{1}{57}$ and $4 \arctan \frac{1}{239}$, all they had to do was compute $48 \arctan \frac{1}{18}$ and add these values with the appropriate coefficients. It is a good thing that they did check their results, since their initial computation of $24 \arctan \frac{1}{8}$ was found to be in error after the 70,695th digit.

5.3. NUMBER THEORETIC PROBLEMS

Experimenting with numbers and looking for patterns in computations is certainly the origin of many problems in number theory. Since digital computers greatly shorten computations and make feasible computations that could never be done by hand, it is quite natural that they have been used extensively in examining unsolved number theoretic problems. Computers are particularly good at pointing out counterexamples that demonstrate the falsity of a conjectured result. For example, Euler (ca. 1769) conjectured that it was impossible to find positive integer values satisfying

$$a^5 + b^5 + c^5 + d^5 = e^5.$$

Recently a CDC 6600 discovered, after an exhaustive search, that

$$27^5 + 84^5 + 110^5 + 133^5 = 144^5,$$

disproving Euler's conjecture.

Of course, not all exhaustive searches end successfully by finding counterexamples, but after such a search one can report that the conjecture being examined has been verified up to some large N. If N is astronomically large, one can say that the conjecture is now "more plausible" than ever. No measure of plausibility has ever been suggested; so the idea of the conjecture being "more" plausible may not be very meaningful.

Number theory abounds with hard problems, and in this section we consider some of these problems and the techniques used to attack them with the computer. Sieves, used to study many problems, are discussed first; next we describe the techniques by which very large primes have been computed, and then we examine a difficult (and yet unsolved) problem to show how insight can be gained through extensive computation.

5.3.1. Sieves

A *sieve*, as the name suggests, is a combinatorial programming technique which takes a finite set and eliminates all the members of that set which are *not* of interest; it is the logical complement of the backtrack process described in Section 2.1, which enumerates all the members of a set which *are* of interest. We discuss sieves here, rather than in Chapter 2, because they are largely of use in number theoretic computations. For example, the sieve of Eratosthenes is one of the best-known methods of finding prime numbers; this sieve enumerates the composite (nonprime) numbers between N and N^2 for some N, as illustrated in Figure 5.3 for $N = 6$. The sieve begins by writing down the integers from N to N^2, and then removing the composite numbers in stages. First, all multiples of 2 are removed; then all multiples of 3 are removed; and so on. The process stops after sifting with largest prime less than N.

We can interpret the sieve of Eratosthenes as searching for all numbers between N and N^2 that are simultaneously members of one of each of the following sets of arithmetic progressions:

$$\{2k + 1\},$$
$$\{3k + 1, 3k + 2\},$$
$$\{5k + 1, 5k + 2, 5k + 3, 5k + 4\},$$
$$\{7k + 1, \ldots, 7k + 6\},$$
$$\{11k + 1, \ldots, 11k + 10\},$$
$$\cdot$$
$$\cdot$$
$$\cdot$$
$$\{pk + 1, \ldots, pk + p - 1\},$$

where p is the largest prime less than or equal to N. The fact that a number is in the progression $2k + 1$ means it is not even; the fact that it is in one of the progressions $3k + 1$ or $3k + 2$ means that it is not a multiple of 3, and so on; so the only numbers between N and N^2 satisfying all these conditions are the primes.

Consider the famous medieval puzzle of the old woman and the eggs. On her way to the market to sell her eggs, a horseman knocks her down by

Stage (0) (initially)

6 7 8 9 10 11 12 13 14 15 16 17 18 19 20 21 22 23 24 25 26 27 28 29 30 31 32 33 34 35 36

Stage 1 (multiples of 2 eliminated)

6̸ 7 8̸ 9 1̸0 11 1̸2 13 1̸4 15 1̸6 17 1̸8 19 2̸0 21 2̸2 23 2̸4 25 2̸6 27 2̸8 29 3̸0 31 3̸2 33 3̸4 35 3̸6

Stage 2 (multiples of 3 eliminated)

6̸ 7 8̸ 9̸ 1̸0 11 1̸2 13 1̸4 1̸5 1̸6 17 1̸8 19 2̸0 2̸1 2̸2 23 2̸4 25 2̸6 2̸7 2̸8 29 3̸0 31 3̸2 3̸3 3̸4 35 3̸6

Stage 3 (multiples of 5 eliminated)

6̸ 7 8̸ 9̸ 1̸0 11 1̸2 13 1̸4 1̸5 1̸6 17 1̸8 19 2̸0 2̸1 2̸2 23 2̸4 2̸5 2̸6 2̸7 2̸8 29 3̸0 31 3̸2 3̸3 3̸4 3̸5 3̸6

Fig. 5.3 The sieve of Eratosthenes used to find all primes between 6 and 36: 7, 11, 13, 17, 19, 23, 29, 31.

accident, causing all the eggs to be broken. He offers to pay the damages and wants to know how many eggs she had. She says she doesn't remember the exact number, but when she took them 2 at a time there was 1 left over. Also, there was 1 left over when they were taken 3, 4, 5, and 6 at a time, but when they were taken 7 at a time it came out even. What are the possibilities for the number of eggs she had? Clearly, she could have had N eggs if and only if N was simultaneously a member of each of the arithmetical progressions

$$2k + 1, \quad 3k + 1, \quad 4k + 1, \quad 5k + 1, \quad 6k + 1, \quad \text{and} \quad 7k.$$

A sieve can be used to solve this problem listing the integers $1, 2, 3, 4, \ldots,$ and then removing all elements not in the various progressions: first all the even numbers, then numbers of the form $3k$ or $3k + 2$, and so on. The numbers that remain are the possible solutions.

This second problem is not unlike the interpretation of the sieve of Eratosthenes given previously, and we can generalize the two of them. Let $m_1, m_2, m_3, \ldots, m_t$ be a set of t positive integers (called the *moduli*). For each m_i we consider n_i arithmetical progressions,

$$m_i k + a_{ij} \begin{cases} i = 1, 2, \ldots, t, \\ j = 1, 2, \ldots, n_i. \end{cases} \tag{1}$$

The problem is now to find all integers, between given limits, that simultaneously satisfy, for each m_i, one of the n_i progressions. In the sieve of Eratosthenes we have $m_1 = 2, m_2 = 3, m_3 = 5, \ldots, m_t = p$ (the largest prime less than or equal to N), $n_i = m_i - 1$, and $a_{ij} = j$. In the problem of the old woman and the eggs we have $m_i = i + 1, 1 \leq i \leq 7, n_i = 1$, and $a_{i1} = 1$, $1 \leq i \leq 6, a_{7,1} = 0$.

Using the solution to Exercise 16, it is clear that given the sequence of t sets of arithmetical progressions in (1), we can combine them into a single set of $n = n_1 n_2 \ldots n_t$ progressions, each of whose modulus is $m = m_1 m_2 \ldots m_t$,

$$mk + b_i, \quad i = 1, 2, \ldots, n.$$

Thus to determine whether a number satisfies one of the n_i progressions for each m_i, we need only divide it by m and see if the remainder is one of $b_1, b_2, \ldots, b_n$. Of course, the size of this new set of progressions grows very rapidly, so rapidly in fact that having only 10 sets of 3 progressions each yields a single set of $n = 3^{10} = 59,049$ progressions. Thus we would have to compute these 59,049 values of b_i and then search them for the remainder we calculated. This is quite a lot of work, and it is usually more practical to examine all the integers between two limits, A and B, excluding each integer in turn for nonmembership in one of the original sets of progressions.

There are many sieves in which the moduli $m_1, m_2, \ldots$ are not predetermined; the value of m_i will depend on the numbers *not* eliminated after sifting with m_{i-1}. Many sieves are constructed in this recursive manner; in fact, the sieve of Eratosthenes is usually so constructed: after writing down the integers $2, 3, \ldots, N$, we cancel all multiples of 2, except 2. Then, since the smallest remaining number whose multiples have not been removed is 3, all multiples of 3, except 3, are removed, and so on. Notice that at each stage the first number removed is the square of the sifting number, and thus the first number eliminated by 2 is 4, by 3, 9, and so on. When the sifting number becomes larger than $\sqrt{N}$, no other numbers can be removed, and so the process ends.

Another well-known recursive sieve produces the *lucky numbers*. From the list of numbers $1, 2, 3, 4, 5, \ldots$ remove every second number leaving the list $1, 3, 5, 7, 9, \ldots$. Since 3 is the first number (excepting 1) that has not been used as a sifting number, we remove every third number *from those remaining numbers*, yielding $1, 3, 7, 9, 13, 15, 19, 21. \ldots$ Now, every seventh number is removed leaving $1, 3, 7, 9, 13, 15, 21, \ldots$. Numbers that are never removed from the list are considered to be "lucky." It is interesting to note that the lucky numbers have the same asymptotic distribution as the prime numbers; that is, they also satisfy the prime number theorem (see Section 5.3.2).

To demonstrate the usefulness of sieves, consider the problem of testing the first one million Fibonacci numbers, as defined in Section 5.1.3, to see which are squares. Since, as we showed in that section,

$$F_n \approx \left(\frac{1 + \sqrt{5}}{2} \right)^n,$$

we know that

$$F_{1,000,000} > 10^{200,000};$$

that is, $F_{1,000,000}$ has over 200,000 decimal digits. It is quite obvious that the brute-force method of generating the first one million Fibonacci numbers and testing each one to see if it is a square is absolutely unfeasible, but the computation can be easily done in a few minutes with a sieve technique.

The calculation begins by setting aside one million bits in memory. The ith bit will represent the integer F_i; if that bit is 1 then F_i *might* be a square, if it is 0 then F_i cannot be a square. Initially, all the bits will be 1, and during the sifting process certain of these bits are set to 0 at each stage. When the sifting is completed, if any bit position is 1, the corresponding Fibonacci number must be examined to see if it is a square. The one million bits of memory required by the sieve is not excessive; for example, on an IBM 360 this is less than 32,000 words of storage.

Consider the Fibonacci sequence modulo p for a prime p. Let this sequence be $P_1, P_2, P_3, \ldots$ defined by

$$P_1 = P_2 = 1,$$
$$P_{i+1} = P_{i+1} + P_i \qquad (\text{modulo } p).$$

This sequence is periodic, the period starting with $P_1 = 1$. To prove this, consider the p^2 pairs of integers

$$(P_1, P_2), (P_2, P_3), \ldots, (P_{p^2}, P_{p^2+1}).$$

If these pairs are all distinct, one of them must be $(0, 1)$ since there are only p^2 distinct such pairs, each of which must appear in the list. Then we have $(P_t, P_{t+1}) = (0, 1)$, and so $(P_{t+1}, P_{t+2}) = (1, 1) = (P_1, P_2)$, and the sequence has a period of t starting at P_1, so that $P_i = P_{i+nt}$ for $n \geq 0$. On the other hand, if two of the pairs are equal, say

$$(P_r, P_{r+1}) = (P_s, P_{s+1}), \qquad r < s,$$

then we have

$$(P_{r-1}, P_r) = (P_{s-1}, P_s),$$

and by induction

$$(P_1, P_2) = (P_{s-(r-1)}, P_{s+1-(r-1)});$$

so the sequence has a period of $s - r$ starting at P_1, so that $P_i = P_{i+n(s-r)}$ for $n \geq 0$.

Now modulo any number n there are certain squares and certain non-squares. For example, 0, 1, 2, and 4 are squares modulo 7, whereas 3, 5, and 6 are nonsquares. Thus any number $m \equiv 3, 5,$ or 6 (mod 7) cannot be a square, since it is not a square modulo 7. Since for every prime p the Fibonacci numbers are periodic modulo p, we sift using the arithmetical progressions $pk + a_i$, for each a_i that is a nonsquare modulo p. To illustrate the sieve, consider $p = 7$. The Fibonacci series modulo 7 is the sequence of 16 elements

$$1, 1, 2, 3, 5, 1, 6, 0, 6, 6, 5, 4, 2, 6, 1, 0$$

repeated over and over again. Since 3, 5, and 6 are nonsquares, we know that $F_{4+16n}, F_{5+16n}, F_{7+16n}, F_{9+16n}, F_{10+16n}, F_{11+16n},$ and F_{14+16n} are not squares for $n \geq 1$, and so the bits corresponding to those Fibonacci numbers can be changed from 1 to 0.

After sifting with the first 32 prime numbers, we find that all the bits are changed to zeroes except the first, second, and twelfth. These correspond to the three numbers $F_1 = F_2 = 1$ and $F_{12} = 144$, and so we can conclude that these are the *only* squares among the first 1 million Fibonacci numbers. In fact, it has been proved that these are the only square Fibonacci numbers.

5.3.2.   Large Primes

The history of mathematics is rich in conjectures about prime numbers, and some of these have had a lasting influence on the development of mathematics. The main reason that the primes are such a fertile ground for influential conjectures is that one can state questions which are easily understood by anyone with a high school exposure to mathematics, yet which are so difficult as to defy solution for centuries.

Many conjectures about primes appear to have originated by observing patterns that appeared in calculations. One famous example concerns the distribution of primes. If we count the number of primes in sufficiently long intervals, we observe a smooth, regular behavior. Let $\pi(x)$ denote the number of primes that are less than or equal to x. Gauss conjectured in 1792 that for large x, $\pi(x)$ behaves asymptotically like $x/\ln x$. It took more than 100 years before Hadamard and de la Vallée Poussin proved this now famous *prime number theorem*.

The hope of observing patterns that serve as a guide to interesting conjectures and proofs remains a powerful incentive to undertake extensive calculations. The other main incentive, which will become apparent in the following story of the search for large primes, is gamesmanship—people are so proud to be able to write down a prime number explicitly which is larger than any prime written before that they have even spread the news by putting the message on business envelopes (see Figure 5.4)! Toward the end of this section we shall discuss briefly what sorts of checks and precautions two respectable institutions might have undertaken before staking their reputation on the accuracy of some rather lengthy calculations; but let us start with the intriguing story of the race for the largest known prime.

As Euclid proved millenia ago, there are infinitely many primes, and so this race can keep people who have the right frame of mind busy and happy for a long time to come. Since there are several straightforward ways of generating larger and larger primes, it is not immediately apparent why anybody would be interested in actually writing down a large prime. The reason is that the straightforward techniques do not allow us to go very far, within the limits of what is computationally feasible. The most obvious technique is to check each natural number in turn by dividing it by all primes smaller than the square root of this number; another one is the sieve technique of Section 5.3.1.

Although both techniques may be improved in a number of ways, neither would enable us to compete seriously in the race for the largest known prime. A remark by Mersenne, the seventeenth century mathematician who made

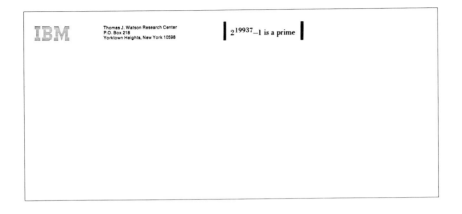

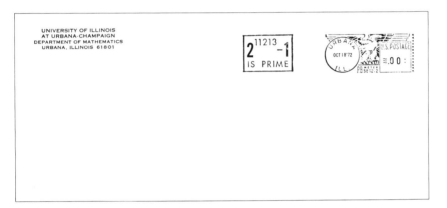

Fig. 5.4 Samples of business envelopes used to herald the news of the discovery of new primes.

a prominent contribution to today's race for large primes, serves to make the point: "To tell if a given number of 15 or 20 digits is prime or not, all time would not suffice for the test, whatever use is made of what is already known." Mersenne was undoubtedly thinking of hand calculations, but today's high-speed computers would still not suffice to reach a number of the size of $2^{19,973} - 1$ by using straightforward methods (see Exercise 19).

The largest primes known are of the form $2^p - 1$, where p is itself a prime. They are called Mersenne primes, since Mersenne brought these numbers into prominence by claiming that $p = 2, 3, 5, 7, 13, 17, 19, 31, 67, 127, 257$ give rise to all and only primes of the form $2^p - 1$ with $p \leq 257$. Mersenne made errors of commission as well as omission in this list (67 and 257 should be deleted; 61, 89, 107 should be added). He knew what he was talking about when he said all time would not suffice to test a 20-digit number for primality!

What makes it possible to check large numbers of the form $2^p - 1$ for primality is the following *Lucas test*:

Theorem

If $p > 2$ is a prime, then $2^p - 1$ is prime if and only if $L_{p-2} = 0$, where the sequence L_i is defined as follows:

$$L_0 = 4, \quad L_{i+1} = (L_i^2 - 2) \text{ modulo } (2^p - 1).$$

Lucas's test, published in 1878, allowed the determination by hand of all Mersenne primes up to $2^{127} - 1$. When computers were applied to the task, the race really started. The following new Mersenne pri es M_p are a testimony to the rapid increase in the power of computers:

$p = 521, 607, 1279, 2203, 2281$ (R. M. Robinson, 1954),

$p = 3217$ (H. Riesel, 1958),

$p = 5253, 4423$ (A. Hurwitz and J. L. Selfridge, 1962),

$p = 9689, 9941, 11,213$ (D. Gillies, 1964),

$p = 19937$ (B. Tuckerman, 1971).

A comparison of actual computation times gives further indications of the improvement in the speed of computers, which enabled later investigators to reach higher values of p: to determine that $2^{8191} - 1$ is *not* a prime (and thus to destroy an earlier conjecture that $2^p - 1$ is always prime if p is a Mersenne prime), ILLIAC I, a fast computer of 1950 vintage took 100 hours (D. J. Wheeler). The same calculation took 5 hours on an IBM 7090 (Hurwitz, 1962), 49 minutes on ILLIAC II (Gillies, 1964), and 3 minutes on an IBM 360 Model 91 (Tuckerman, 1971). The test that $2^{11,213} - 1$ is prime took 2 hours and 15 minutes on ILLIAC II, but only 8 minutes on the IBM 360/91.

There are large, irregularly spaced gaps in the sequence of Mersenne primes, a fact that explains why Mersenne primes have frequently been discovered in clusters. These large gaps have raised the question of how Mersenne primes are distributed. Gillies conjectures that the number of Mersenne primes M_p less than a given number x is asymptotically equal to $(2/\log 2) \log \log x$. No way is in sight of how one might prove this; but then, when Gauss conjectured the prime number theorem, he did not know how to prove it either.

The various computations that have tested Mersenne numbers for primality have each taken on the order of 100 hours of machine time. One should never assume that a computer will run free of errors for such a long time—indeed, many computers have had a fairly short mean time between failures. So, how do we get to have any confidence in the result of a computation whose length is comparable to the mean time between failures of the machine on which it is run?

This is a difficult question, and we can make only a few unsystematic remarks about it; this is indicative of the current state of the art, where lengthy calculations are commonly checked by ad hoc techniques, as should already be clear from Section 5.2. The purpose of discussing this topic will have been served if it helps to spread the idea that one should not necessarily believe something just because it was printed out by a computer.

Probably the first check one would undertake is to compare the results of a new program on the already known cases. Agreement will increase the plausibility that the already known results are indeed correct, as well as that the new program is correct. It will not, however, substantially increase confidence that the computer was actually running correctly while it produced *new* results. To safeguard against the latter errors, redundancy must be included in the computation; that is, quantities must be computed in different ways, and it must be verified that they agree. We have already seen this in Section 5.2; for another example, let us quote from Tuckerman's article (see Section 5.4):

> As a control both on the programming and on the computer operation, each recursion was checked by residues, both mod $2^{24} - 1$ and mod $2^{24} - 3$. The first is very convenient, but it would not detect, for example, the addition of a correct quantity into a wrong digit position. The second check avoids this weakness. It requires more computation than the first check, but the penalty is tolerable. These modular checks never reported a computer error, and did detect synthetic errors. Consequently, I felt it was sufficient to run most of the cases of p only once.
>
> This program . . . was run first on the lower part of the known range of p, for confirmation, and then into the unknown range. Production was done chiefly during otherwise idle times at the end of the third shift. On the evening of March 4, 1971, a zero Lucas–Lehmer residue for $p = p_{24} = 19{,}937$ was found. Hence, $M_{19,937}$ is the 24th Mersenne prime.

Tuckerman certainly employed the techniques that are customary today to check a computation. Yet, when he says "Hence $M_{19,937}$ is the 24th Mersenne prime," one cannot help but wonder what the probability is that this statement is wrong. It is important that we retain a certain skepticism toward the answers computers give us.

5.3.3. Multiply by 3 and Add 1

It is easy to ask difficult questions about properties of numbers. In this section we shall consider various questions that are suggested by the following algorithm whose input is a natural number k:

> If $k = 1$, stop. If k is even, set $k \leftarrow k/2$; if k is odd, set $k \leftarrow 3k + 1$.
> Repeat, using the new value of k.

For example, if we start with $k = 34$, the algorithm generates the sequence

$$34, 17, 52, 26, 13, 40, 20, 10, 5, 16, 8, 4, 2, 1.$$

The most natural question concerning this algorithm is whether or not it will halt for every natural number. We do not know the answer to this question; we can, however, observe many interesting patterns that would have been difficult to discover without fairly extensive computation. Let S be the function defined by

$S(k)$ = number of steps the algorithm will run given an initial value k.

The question now becomes whether or not S is defined for all natural numbers.

Table 5.3 gives values of $S(k)$ for the natural numbers up to 50. In addition to the value of $S(k)$, this table gives, for each k, the highest number generated by the algorithm and the first power of 2 generated by the algorithm. Computing this other information is an attempt to establish a guide for future classification of the natural numbers into classes that might make patterns easier to discover.

A few interesting observations are immediate. First, in nearly all cases, the first power of 2 generated is 16. At first this is quite striking, but a little thought makes it understandable. It is clear that the first power of 2 will be generated by the multiply-by-3-and-add-1 part of the algorithm so that it is of the form $3m + 1$ for some odd number m, or, equivalently, $6m + 4$ for some natural number m. Looking at the powers of 2, we see that this will be true only for the powers of 2 with even exponent, that is, 4, 16, 64, 256, and so on; so no odd power of 2 will appear in this column more than once. This explains the absence of 32 in this column (except for $k = 32$). The relative scarcity of 64 is still puzzling but will be explained later.

A few numbers occur frequently as the maximum number generated, most notably 52, 88, 160, and 9232, together accounting for half the sequences between 1 and 50.

Looking for patterns among the maximum numbers generated yields this interesting fact: all of them have at most one prime factor different from 2. Unfortunately, this pattern is quickly destroyed by a maximum of 340 for $k = 75$. The short initial segment ($k \leq 50$) of the complete table is misleading in this respect, but extending the table to $k \leq 100$ does not change the fact that relatively few numbers have a tendency to occur frequently in

Table 5.3 SOME DATA COLLECTED FROM THE ALGORITHM.

k	S(k)	Max Number Generated	First Power of 2 Generated
1	0	1	1
2	1	2	2
3	7	16	16
4	2	4	4
5	5	16	16
6	8	16	16
7	16	52	16
8	3	8	8
9	19	52	16
10	6	16	16
11	14	52	16
12	9	16	16
13	9	40	16
14	17	52	16
15	17	160	16
16	4	16	16
17	12	52	16
18	20	52	16
19	20	88	16
20	7	20	16
21	7	64	64
22	15	52	16
23	15	160	16
24	10	24	16
25	23	88	16
26	10	40	16
27	111	9232	16
28	18	52	16
29	18	88	16
30	18	160	16
31	106	9232	16
32	5	32	32
33	26	100	16
34	13	52	16
35	13	160	16
36	21	52	16
37	21	112	16
38	21	88	16
39	34	304	16
40	8	40	16
41	109	9232	16
42	8	64	64
43	29	196	16
44	16	52	16
45	16	136	16
46	16	160	16
47	104	9232	16
48	11	48	16
49	24	148	16
50	24	88	16

the column of maxima. With a few remarkable exceptions [which have $S(k)$ > 100] all the first 50 numbers have a sequence length $S(k) < 35$.

A final observation is that there are several sequences of consecutive numbers which have the same value. For example, $S(14) = S(15) = 17$, $S(28) = S(29) = S(30) = 18$,   and $S(98) = S(99) = S(100) = S(101) = S(102) = 25$. To illustrate, consider the sequences for the pair 14, 15:

$$14 \quad 7 \ 22 \ 11 \ 34 \quad 17 \ 52 \quad 26 \ 13 \ 40 \ 20 \ 10 \ 5 \ 16 \ 8 \ 4 \ 2 \ 1$$
$$15 \ 46 \ 23 \ 70 \ 35 \ 106 \ 53 \ 160 \ 80 \ 40 \ 20 \ 10 \ 5 \ 16 \ 8 \ 4 \ 2 \ 1$$

Notice the regular connection between numbers in corresponding positions in each sequence. This phenomenon suggests a second approach to the problem, that of looking for patterns in the classes of numbers which have the same sequence length s.

Let $C(s)$ denote the set of numbers for which the algorithm runs for s steps. That is, a number k is in $C(s)$ if and only if $S(k) = s$. For example, $C(0) = \{1\}$, $C(1) = \{2\}, \ldots, C(5) = \{32, 5\}$. The computation of $C(s)$ for $s = 0, 1, 2, \ldots$ is best accomplished by a recursive definition:

$C(0) = \{1\}$.
Given $C(s) = \{k_1, k_2, \ldots, k_M\}$, $C(s + 1)$ will contain exactly the following numbers
    1. For each k_i, $2k_i$.
    2. For each k_i of the form $6n + 4$, $(k_i - 1)/3$.

Executing this algorithm for $s = 1, 2, \ldots$ has the effect of building a tree, a small part of which is illustrated in Figure 5.5. The horizontal rows in the tree consist of exactly the classes $C(s)$, $s = 0, 1, 2, \ldots$ and a path from a node of the tree to the root follows the sequence of numbers generated by the algorithm.

This picture explains why 64 appears so infrequently in Table 5.3. In order for 64 to be the first power of 2 generated for any number k, k must be one of the numbers in the branch of the tree above 21. Since 3 divides 21, it is clear that none of these numbers will be of the form $6n + 4$. Hence by the previous recursive definition, the numbers for which 64 is the first power of 2 generated are exactly those of the form $21 \cdot 2^m$ for some m.

The recursive algorithm for computing the classes $C(s)$ leads to a recurrence relation for computing $n(s)$, the number of elements in $C(s)$. Let $e(s)$ and $o(s)$ denote the number of even and odd elements of $C(s)$, respectively, so that $n(s) = e(s) + o(s)$. Now notice that for each s, $e(s) = n(s - 1)$ (Why?) and that every odd number in $C(s)$ arises from a number of the form $6n + 4$ in $C(s - 1)$.

Let us assume that roughly one third of all even numbers in any $C(s)$

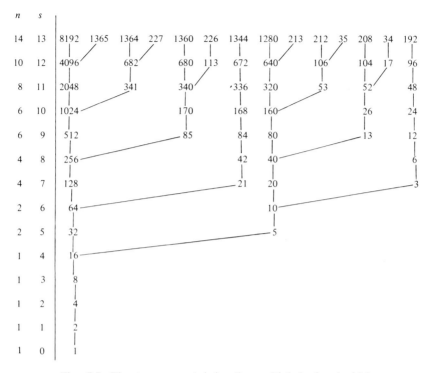

Fig. 5.5 The tree generated by the multiply-by-3-and-add-1 algorithm.

are of the form $6n + 4$. [This is not an obvious fact. For example, from the tree in Figure 5.5 it can be seen that in any $C(s)$ there are systematically fewer odd numbers than even ones.] Then we get the relation

$$o(s) = \frac{e(s-1)}{3},$$

and since

$$e(s-1) = n(s-2),$$

we obtain the recurrence relation

$$n(s) = n(s-1) + \frac{n(s-2)}{3},$$

which has a solution of the form

$$n(s) = \text{constant} \cdot \left(\frac{1+\sqrt{7/3}}{2}\right)^s,$$

so, $n(s)$ grows exponentially (see Section 5.1.3); the base

$$\frac{1 + \sqrt{7/3}}{2} \approx 1.26$$

is nicely confirmed by observing the ratios $n(s)/n(s-1)$ in Table 5.4.

Table 5.4 GROWTH OF $n(s)$.

s	$n(s)$	$e(s)$	$e(s)/n(s)$	$n(s)/n(s-1)$
0	1	0	0.00	
1	1	1	1.00	1.00
2	1	1	1.00	1.00
3	1	1	1.00	1.00
4	1	1	1.00	1.00
5	2	1	0.50	2.00
6	2	2	1.00	1.00
7	4	2	0.50	2.00
8	4	4	1.00	1.00
9	6	4	0.67	1.50
10	6	6	1.00	1.00
11	8	6	0.75	1.33
12	10	8	0.80	1.25
13	14	10	0.71	1.40
14	18	14	0.78	1.29
15	24	18	0.75	1.33
16	29	24	0.83	1.21
17	36	29	0.81	1.24
18	44	36	0.82	1.22
19	58	44	0.76	1.32
20	72	58	0.81	1.24
21	91	72	0.79	1.26
22	113	91	0.81	1.24
23	143	113	0.79	1.27
24	179	143	0.80	1.25
25	227	179	0.79	1.27

There are many interesting patterns to be observed in the classes $C(s)$. For example, consider the class $C(16)$ shown in Table 5.5. A glance at the largest numbers in this class reveals that differences between them are often powers of 2, suggesting that we write the numbers in base 2. This makes the pattern in the numbers from 10,240 to 10,922 very apparent. A similar pattern exists in every class $C(s)$; this is explained by considering what happens to a number whose binary representation has the form

$$101010\ldots101000\ldots0$$

when subjected to the algorithm.

Table 5.5   ELEMENTS OF $C(16)$ WRITTEN IN BOTH
DECIMAL AND BINARY NOTATIONS.

In Decimal	In Binary
65,536	10000000000000000
10,922	10101010101010
10,920	10101010101000
10,912	10101010100000
10,880	10101010000000
10,752	10101000000000
10,240	10100000000000
1,818	11100011010
1,816	11100011000
1,813	11100010101
1,812	11100010100
1,808	11100010000
1,706	11010101010
1,704	11010101000
1,696	11010100000
1,664	11010000000
1,536	11000000000
302	100101110
301	100101101
300	100101100
282	100011010
280	100011000
277	100010101
276	100010100
272	100010000
46	101110
45	101101
44	101100
7	111

5.4. REMARKS AND REFERENCES

All the topics in Section 5.1 are discussed in elementary numerical analysis texts. We have attempted to present only a few basic principles and techniques; complete analyses can be found, for example, in

> FRÖBERG, C. E. *Introduction to Numerical Analysis*, Addison-Wesley,
> Reading, Mass., 1965,

or in

> RALSTON, A. *A First Course in Numerical Analysis*, McGraw-Hill,
> New York, 1965.

A thorough discussion of the floating-point representation of real numbers and of the errors in computer arithmetic operations is given in

WILKINSON, J. H. *Rounding Errors in Algebraic Processes*, Prentice-Hall, Englewood Cliffs, N.J., 1963.

For additional examples and problems see

ANDREE, R. V., J. P. ANDREE, and D. D. ANDREE. *Computer Programming: Techniques, Analysis, and Mathematics*, Prentice-Hall, 1973.

The calculation of mathematical constants to ultrahigh precision is really just an art, since there is no general body of information and there are no general principles. We have presented some of the ad hoc techniques that have been developed; the complete details of these computations are to be found largely in unpublished technical reports and occasionally in journal articles.

The calculation of $\sqrt{2}$ and its history are described in

DUTKA, J. "The Square Root of 2 to 1,000,000 Decimals," *Math. Comp.*, *25* (1971), 927–930.

Other references can be found by looking at those cited by Dutka and those cited in the papers which Dutka cites. It is interesting to note that the approximation of $\sqrt{N}$ by the iteration

$$x_{k+1} = \frac{1}{2}\left(x_k + \frac{N}{x_k}\right)$$

(Newton's method) dates back to Heron the Elder, about 100 B.C.!

The major calculations of e by Wheeler, Shanks and Wrench, and Gillies and Wheeler have never been published. It is only through footnotes and personal communication that these computations are known to exist.

Unlike $\sqrt{2}$ and e, there is a considerable body of literature devoted to the calculation of π. A brief historical account and a proof of the transcendence of π (and e) is given in

SMITH, D. E. "The History and Transcendence of π," in *Monographs on Topics of Modern Mathematics*, J. W. A. Young (ed.), Dover, New York, 1955.

A detailed history and bibliography of the calculations of π up to 1960 is

WRENCH, J. W. "The Evolution of Extended Decimal Approximations to π," *Math. Teacher*, *53* (1960), 644–665.

The calculation of π to 250,000 and 500,000 places has never been published, but the value of π to 100,000 places and its calculation are given in

SHANKS, D., and J. W. WRENCH. "Calculation of π to 100,000 Decimals," *Math. Comp.*, *77* (1962), 76–99.

A popularized account of the history of π is

BECKMANN, P. *A History of π*, 2nd ed., Golem Press, Boulder, Colo., 1971.

The use of computers in number theoretic problems is discussed in

LEHMER, D. H. "Machines and Pure Mathematics," in *Computers in Mathematical Research*, R. F. Churchhouse and J. C. Herz (eds.), North-Holland, Amsterdam, 1968.

Lehmer is a pioneer in the field, and he gives several examples of problems that have been solved, or partially solved, with the aid of a computer. That same book contains an extensive bibliography of work in the area.

Lehmer is also a pioneer in the use of sieves on computers. His classical paper,

LEHMER, D. H. "The Sieve Problem for All-Purpose Computers," *Math. Tables Aids Comput.*, 7 (1953), 6–14,

contains a cogent summary of the generalized modular sieve. Various sieves and their properties and uses are discussed in

HAWKINS, D. "Mathematical Sieves," *Sci. Amer.*, *199*, No. 6 (Dec. 1958), 105–112,

and in

WUNDERLICH, M. C. "Sieving Procedures on a Digital Computer," *J. ACM*, *14* (1967), 10–19.

The sieve to determine perfect-square Fibonacci numbers is presented in

WUNDERLICH, M. C. "On the Existence of Fibonacci Squares," *Math. Comp.*, *17* (1963), 455–457.

The proof that 1 and 144 are the only such squares can be found in

COHN, J. H. E. "Square Fibonacci Numbers," *Fibonacci Quart.*, *2* (1964), 109–113.

A detailed discussion of many techniques for computing with primes can be found on pages 338–359 of

KNUTH, D. E. *The Art of Computer Programming*, Vol. 2, Addison-Wesley, Reading, Mass., 1969.

The two most recent advances toward larger Mersenne primes are described in the papers

GILLIES, D. B. "Three New Mersenne Primes, and a Statistical Theory," *Math. Comp.*, *18* (1964), 93–97,

and

TUCKERMAN, B. "The 24-th Mersenne Prime," *Proc. Natl. Acad. Sci. U.S.*, *68* (1971), 2319–2320.

The multiply-by-3-and-add-1 problem is still unsolved despite the fact that number theorists have worked quite hard on it.

5.5. EXERCISES

1. Prove that for t digit floating-point arithmetic the computed value of $a \circ b$ is equal to the exact value of $a \circ b$ times $(1 + \epsilon), |\epsilon| \leq 0.5 \times 10^{1-t}$, where $\circ$ is addition, subtraction, multiplication, or division.

2. Calculate the "Manhattan problem" ($24 invested at 3 percent per year for 300 years) year by year in the following different ways:
   (a) *Exactly*—every digit carried at every stage.
   (b) Using integer arithmetic *rounded* to the nearest cent after each iteration.
   (c) Using integer arithmetic *truncated* to the nearest cent after each iteration.
   (d) Using 12-digit floating-point arithmetic.
   Then do the calculation using the compound interest formula and 6-digit floating-point arithmetic. Discuss the differences between these answers.

3. Compute $\sum_{i=1}^{1000} 1/i$ and $\sum_{i=1}^{1000} (-1)^i/i$ from left to right and from right to left. Discuss the differences between corresponding results.

4. How many positive numbers can be represented by the floating-point notation used in Section 5.1.1? How are these numbers distributed between 0 and 0.9999×10^{99}?

5. Let f be an integrable function of one real variable. For a given interval (a, b) define a sequence S_n by

$$S_n = \left(\frac{b - a}{2^n}\right) \sum_{i=1}^{2^n} f\left(a + \frac{i(b - a)}{2^n}\right);$$

thus S_n is an approximation to the integral $\int_a^b f(x)\, dx$ using the rectangle rule with 2^n equal subdivisions of the interval (a, b). Assume that f is differentiable at least once on the interval (a, b) and that M is a bound for the first derivative of f on the interval (a, b).
   (a) Derive an estimate of the error

$$\left| \int_a^b f(x)\, dx - S_n \right|.$$

   How does it depend on n?
   (b) What is the order of convergence of the sequence S_n?

6. (a) Apply the method of Section 5.1.3 to find a closed-form expression for the recurrence relation

$$F_{k+3} = -3F_{k+2} - F_{k+1} + F_k.$$

   (b) Let A_k and B_k be sequences satisfying the above recurrence relation and having initial values

$$A_0 = 2, \qquad A_1 = -2 + \sqrt{2}, \qquad A_2 = 4 - 2\sqrt{2},$$
$$B_0 = -1, \qquad B_1 = 1 - 4\sqrt{2}, \qquad B_2 = -5 + 8\sqrt{2}.$$

Determine the coefficients in the closed form found in (a) that give closed-form expressions for A_k and B_k. Conclude that the sequence A_k cannot be accurately computed using the recurrence relation; that is, such a computation is unstable.

(c) Verify the conclusion of (b) with a program that computes the first 24 terms of A_k, using both the closed-form and the recurrence relation. Compute also, using the recurrence relation, the sequences C_k, D_k, E_k, F_k with initial values (approximating A_k) of

$$C_0 = 2, \qquad C_1 = -0.5, \qquad C_2 = 1.1,$$
$$D_0 = 2, \qquad D_1 = -0.58, \qquad D_2 = 1.17,$$
$$E_0 = 2, \qquad E_1 = -0.585, \qquad E_2 = 1.171,$$
$$F_0 = 2, \qquad F_1 = -0.5857, \qquad F_2 = 1.1715.$$

Use the results of these computations to discuss how the instability is influenced by the size of the error in the initial values of the sequence. Can you find a simple relationship between the number of correct decimal digits in the initial values and the point in the sequence at which the computation "blows up"?

7. Write routines to add, subtract, multiply, and divide (with remainder) integers of ultra-high precision, say, several thousand digits long.

8. Prove that the limit of the sequence $x_1 = 1$, $x_{k+1} = x_k(\frac{3}{2} - x_k^2)$ converges to $\sqrt{2}/2$.

9. M. Lal used the following digit-by-digit method to compute $\sqrt{2}$ to 19,600 decimal places: Let $A_0 = 1$ and $A_{k+1} = 10A_k + a_{k+1}$ and define

$$B_0 = 1,$$
$$B_{k+1} = 100B_k - \sum_{n=1}^{a_{k+1}} (20A_k + 2n - 1),$$

where $a_0 = 1$ and a_{k+1} is the largest value of n for which $B_{k+1} > 0$; if there is no such value of n, then $a_{k+1} = 0$. Then, for example,

$$A_0 = 1, \qquad B_0 = 1, \qquad a_0 = 1,$$
$$A_1 = 14, \qquad B_1 = 4, \qquad a_1 = 4,$$
$$A_2 = 141, \qquad B_2 = 119, \qquad a_2 = 1,$$

and so on. Prove that $a_0 \cdot a_1 a_2 a_3 a_4 a_5 \ldots$ is the base 10 representation of $\sqrt{2}$.

10. Define $A_1 = 1$ and $A_{k+1} = [\sqrt{2A_k(A_k + 1)}]$, and let $a_k = A_{2k+3} - 2A_{2k+1}$

for $k = 1, 2, \ldots$. Prove that $a_0 \cdot a_1 a_2 a_3 a_4 a_5 \ldots$ is the base 2 representation of $\sqrt{2}$.

Hint: Show that $A_k = [2^{(k-1)/2} + 2^{(k-2)/2}]$.

11. Prove that if $P_0^2 - NQ_0^2 = 4$ and

$$P_{k+1} = P_k^2 - 2,$$
$$Q_{k+1} = P_k Q_k,$$

then $\lim\limits_{k \to \infty} (P_k/Q_k) = \sqrt{N}$.

12. Carry out the first five steps of Dutka's calculation of $\sqrt{2}$. How accurate is your approximation?

13. Show that the way Wheeler checked his computation actually works. Derive a similar method of checking the calculations that extends the precision of $\sum_{i=1}^{n} 1/i!$.

14. Derive Archimedes' upper bound by computing the area in a 96-gon circumscribed about the unit circle.

15. Compute i^i, $\sqrt{3}$, π, e, $\ln 2$, and $\cos 2°$ to 1000 decimal digits of accuracy and use the chi-square test from Section 4.1.2. to determine whether the frequency of occurrence of the digits is random.

16. Prove that any two arithmetical progressions with relatively prime moduli

$$m_1 k + a_1, \qquad m_2 k + a_2$$

can be combined into a single one of the form

$$m_1 m_2 k + a_3,$$

which consists of all numbers common to the original two progressions; for example,

$$5k + 4, \qquad 12k + 5$$

yield

$$60k + 29.$$

17. Write a program to generate all the lucky numbers less than 10,000.

18. Program the sieve discussed in Section 5.3.1 to examine the first 1000 Fibonacci numbers for perfect squares.

19. Mersenne stated that by means of hand calculations it was not feasible to tell if a given number of 15 or 20 digits is prime. Estimate how large a number you could test for primality on some computer you know (in terms of the memory available and a maximal computation time of, say, 1 day):
    (a) By the technique of dividing by smaller primes.
    (b) By the sieve method of Section 5.3.1.

20. Compare the work required to carry out the Lucas test to check whether a number $2^p - 1$ is prime with the work required by a sieve method or by a division method, assuming that you have a binary computer whose word length is w. How would you carry out the operation n mod $(2^p - 1)$? How many multiplications of pairs of numbers of w bits each does the Lucas test require?

21. If 1 is *subtracted* instead of added in the algorithm of Section 5.3.2, does the algorithm always halt?

22. Prove that 1, 2, 145, and 40,585 are the only integers with the property that they are equal to the sum of the factorials of their decimal digit; for example,

$$145 = 1! + 4! + 5!.$$

Hint: Use a computer to check all integers between 1 and 2,000,000; then separately consider integers between 2,000,000 and 2,899,999, between 2,900,000 and 2,999,999, and greater than 3,000,000.

23. Prove that for any k there is some n so that 2^n (in decimal notation) has a string of k zeros. Find the smallest values of n in which 2^n contains strings of k, $1 \leq k \leq 8$.

24. Rumor has it that the multiply-by-3-and-add-1 algorithm has been shown to stop for all values of k up to 10^{40}, so it is usually conjectured that it stops for all k. Using sieve techniques, verify as much of this range as you can.

6 WHAT MACHINES CAN AND CANNOT DO

The question "What can machines do and what can they not do?" has certainly not been raised for the first time in connection with computers. People always appear to have wondered how far the performance of machines typical of the technology of the era could be pushed. In particular, the idea of building machines that simulate human performance in some respect always seems to have exerted a powerful attraction. There were always a few inventors who dreamed, designed, and occasionally even built such machines as a steam-driven warrior or an automatic scribe driven by a clockwork.

However, the great increase in the speed of computation and in the complexity of machines that has occurred over the past three decades and our greatly advanced understanding of how to design algorithms to make machines perform the way we like give to this question a much greater importance today than it ever had before. And, the increasingly pervasive role that computers and computer-controlled machines play in our society makes it important that the general public acquire some understanding of the potential and limitations of machines.

There is no one clear answer to our question "What can a machine do and what can it not do?", if only because the words "can do" are so vague as to allow a variety of interpretations. However, over the past 50 years some rather precise questions relating to the one above have been formulated and partially answered. An awareness and understanding of some of these questions help to avoid much confusion about the capabilities of machines and to dispel some widespread myths.

6.1. CAN A MACHINE THINK?

6.1.1. Turing's Test

The provocative question, "Can a machine think?", is probably one of the more recent questions to be asked about machines. Until the advent of digital computers, machines were designed almost exclusively to perform purely mechanical tasks, and thus provided little motivation for wondering about their intellectual abilities.

There must have been some earlier instances, however, that did raise the question of machine thought in the minds of many people. A case in point is a fake chess-playing automaton, designed by a Baron von Kempelen around 1800, which was exhibited for many years and is reputed to have won all the games it ever played. Inside the machine was a man so cleverly concealed that, although spectators were allowed to view every part of the interior of the machine at different times, he was able to avoid detection by shifting around.

Today there are several chess-playing programs that play at the level of lower-ranked tournament players, and would beat a majority of people who play chess only occasionally. There is a checker-playing program that has reached the master level. Most people would agree that such a performance is a nontrivial intellectual achievement. Discussions of machines' capabilities in performing intellectual tasks, however, have often been clouded by irrational emotional arguments.

The question of thinking machines was discussed in 1950 by the famous logician A. M. Turing, and even today his analysis is one of the best to read in order to clarify one's views of it. Turing accurately sees the question as an emotional one, prone to subjective interpretations of what "thinking" means, and very early dismisses it as being too vague to allow a meaningful answer. Indeed, he points out that for some interpretations the answer to "Can a machine think?" cannot possibly be achieved by logical argument, but is a matter of faith; for others the answer is a foregone conclusion. (Some arguments about thinking machines are based on the hidden assumption that thinking is, by definition, what people do; hence machines cannot think—unless one wants to open the issue as to whether machines are people, or people are machines, which we do not intend to do.)

Turing proposes a new question to replace the original one: a model in which machine "thinking" can be discussed entirely in technical terms. It has since become known as *Turing's test*, and is described as follows. Imagine a computer, C, and a man, M, secluded in a room, each one connected to a teletype in another room where there is a human interrogator, I. I's task is to determine which teletype is connected to C and which to M by typing questions on either or both of the teletypes, and looking at the answers

he receives. This situation then becomes a test for the computer, C, if we give C the task of frustrating I in his effort; that is, C should try to answer I's questions in such a way that I cannot tell C from M. The question then raised by Turing's test, replacing the original "Can a machine think?", is "Are there machines that could be made to do well in the game?".

It is clear that this question has a different character than the original one, for what do we mean by "do well" in the game? It is a matter of degree rather than a simple yes-or-no affair. The criterion for success would likely be statistical: for computer C, on the average, the inquirer, I, has a p per cent chance of choosing correctly after questioning for m minutes. The point is that Turing's test raises issues of engineering design, and in particular of software design: how well can we make a machine to play the game? What can we expect to be able to do 10 or 20 years from now? Disagreement is still possible on such issues, but the emotion that fills the original question has all but disappeared.

In his 1950 paper, Turing stated the belief that by the end of the century machines could be made to perform well enough in the game that an average interrogator would not have more than a 70 per cent chance of making the right identification after 5 minutes of questioning.

We are not aware of anybody having tried directly to make a machine perform in Turing's test. However, a fair amount of work has been done that sheds some light on how one would go about programming a computer for Turing's test, and how well one might expect it to do, if an effort were made with today's state of the art in heuristic programming.

Since a computer, in order to do well in Turing's test, must be made to converse reasonably well via printed words in some natural language, the most relevant insight comes from certain programs that were designed to hold plausible conversations in narrow domains of discourse. We shall discuss two such programs that have been written in recent years. We leave the reader the judge for himself, after seeing these examples, the accuracy of Turing's foresight.

6.1.2. Conversational Programs

Computers have been programmed to hold plausible, and sometimes useful, conversations in limited domains of discourse. There continues to be lively research activity in this area of computer programs that accept and/or respond in natural language.

The purpose of this section is to give the reader some insight into how such programs work, so that he might get an understanding of the relationship between the performance of such a program and its complexity. He might then make his own guess as to how well one could make a program

perform in Turing's test, with a concentrated effort today or 10 years from now.

A program that successfully mimics a good conversationalist should be able to imitate two distinct modes of behavior which occur in human conversation: first, it has to understand a given statement or question in any one of the many forms in which it could have been phrased; that is, it has to get at the meaning of a sentence. Second, it should be able to converse plausibly about a given subject, even if it does not have the facts at hand to give a clear answer or to make an assertion. In other words, it should make a plausible response to such a question as "What do you think of Picasso?".

The distinction between "understanding" and "making a plausible response" is far from clear-cut. However, every reader must have experienced the two states of mind to which we are attempting to draw an analogy: on the one hand, an intense struggle to understand the essence of an argument, and on the other hand, the state of mind when one is not interested in a discussion but, in order to appear polite, nods his head and utters noncommittal phrases like "How interesting."

It is typical of the state of the art that conversational programs which have been written so far tend to emphasize one or the other of these two aspects, but not both. We are still at the stage where researchers try to perfect one technique at a time rather than incorporate several into the same program.

Two programs among several which have been written in recent years, called ELIZA and STUDENT, illustrate the two extremes ("conversation with little understanding," versus "understanding with little conversation") rather well.

ELIZA attempts to carry out a plausible conversation without trying to "understand" the subject under discussion. The word "understanding" is used in the following sense: if a human acted the way we will describe next, we would undoubtedly say that he responded without understanding the issue under discussion.

ELIZA itself is a general-purpose program that accepts sentences in English and responds with sentences of its own, which are obtained from the input sentences according to certain transformations. These transformations are determined by a *script*, and ELIZA behaves differently depending on the particular script it is following. One such script, called DOCTOR, causes ELIZA to respond much the same way a psychiatrist might. It has produced some remarkable conversations, of which the following is an example (ELIZA's responses are printed in capitals).

Men are all alike.
IN WHAT WAY
They're always bugging us about something or other.
CAN YOU THINK OF A SPECIFIC EXAMPLE

Well, my boyfriend made me come here.
YOUR BOYFRIEND MADE YOU COME HERE
He says I'm depressed much of the time.
I AM SORRY TO HEAR YOU ARE DEPRESSED
It's true. I am unhappy.
DO YOU THINK COMING HERE WILL HELP YOU NOT TO
    BE UNHAPPY
I need some help, that much seems certain.
WHAT WOULD IT MEAN TO YOU IF YOU GOT SOME HELP
Perhaps I could learn to get along with my mother.
TELL ME MORE ABOUT YOUR FAMILY
My mother takes care of me.
WHO ELSE IN YOUR FAMILY TAKES CARE OF YOU
My father.
YOUR FATHER
You are like my father in some ways.
WHAT RESEMBLANCE DO YOU SEE
You are not very aggressive but I think you don't want
    me to notice that.
WHAT MAKES YOU THINK I AM NOT VERY AGGRESSIVE
You don't argue with me.
WHY DO YOU THINK I DON'T ARGUE WITH YOU
You are afraid of me.
DOES IT PLEASE YOU TO BELIEVE I AM AFRAID OF YOU
My father is afraid of everybody.
WHAT ELSE COMES TO MIND WHEN YOU THINK OF YOUR
    FATHER
Bullies.
DOES THAT HAVE ANYTHING TO DO WITH THE FACT THAT
    YOUR BOYFRIEND MADE YOU COME HERE

The method by which ELIZA can produce such responses as the ones above is described by its designer, J. Weizenbaum of the Massachusetts Institute of Technology, as follows:

The gross procedure of the program is quite simple; the text is read and inspected for the presence of a keyword. If such a word is found, the sentence is transformed according to a rule associated with the keyword, if not a content-free remark or, under certain conditions, an earlier transformation is retrieved. The text so computed or retrieved is then printed out.

Consider the sentence "I am very unhappy these days". Suppose a foreigner with only a limited knowledge of English but with a very good ear heard that sentence spoken but understood only the first two words "I am". Wishing to appear interested, perhaps even sympathetic, he may reply "How

long have you been very unhappy these days?" What he must have done is to apply a kind of template to the original sentence, one part of which matched the two words "I am" and the remainder isolated the words "very unhappy these days". He must also have a reassembly kit specifically associated with that template, one that specifies that any sentence of the form "I am BLAH" can be transformed to "How long have you been BLAH", independently of the meaning of BLAH. A somewhat more complicated example is given by the sentence "It seems that you hate me". Here the foreigner understands only the words "you" and "me"; i.e., he applies a template that decomposes the sentence into the four parts:

(1) It seems that　(2) you　(3) hate　(4) me

of which only the second and fourth parts are understood. The reassembly rule might then be "What makes you think I hate you"; i.e., it might throw away the first component, translate the two known words ("you" to "I" and "me" to "you") and tack on a stock phrase (What makes you think) to the front of the reconstruction.

The script DOCTOR consists of about a page and a half of decomposition and reassembly rules as described above. It is indeed surprising that such a relatively simple set of rules can produce as credible a conversation as the one recorded earlier. However, this simplicity comes at a cost. In the script discussed, ELIZA had as one of its goals to conceal its lack of understanding. If no decomposition rule applied, an easy response might be "Now let's talk about something else. Tell me about so and so." A phrase such as this, which requires no understanding at all of the preceding discussion, is always applicable in a conversation (provided it is not used too often).

The second program to be discussed, called STUDENT, solves algebra word problems such as

1. "The sum of three numbers is 9. The second number is three more than two times the first number. The third number equals the sum of the first two numbers. Find the three numbers."

2. "Mary is twice as old as Ann was when Mary was as old as Ann is now. If Mary is 24 years old, how old is Ann?".

When the first of these problems is presented, STUDENT prints out

THE FIRST NUMBER IS .5

THE SECOND NUMBER IS 4

THE THIRD NUMBER IS 4.5

and for the second one, it prints

ANN'S AGE IS 18

As far as STUDENT is concerned, the world consists of verbal descriptions of algebraic equations, in particular linear ones. When presented with such a description, STUDENT proceeds to identify the unknowns involved and the relationships that hold between them, and to translate the verbal description into a set of equations in more conventional mathematical notation.

In the first problem, STUDENT identifies the three unknowns f (first number), s (second number), and t (third number), and sets up the system of equations

$$f + s + t = 9,$$
$$s = 3 + 2f,$$
$$t = f + s.$$

Once this stage has been reached, the rest is trivial, since there are well-known techniques for solving systems of linear equations. Our interest in STUDENT is due entirely to its ability to translate sentences from a somewhat restricted, but informal, subset of English into algebraic equations expressed in a formal notation. Our discussion of how STUDENT can do this translation is necessarily simplified, but we hope that the major ideas and techniques will become apparent.

STUDENT classifies the words and phrases in the sentences it is analyzing into three categories called *variables*, *operators*, and *substitutors*.

Substitutors are phrases that STUDENT immediately replaces by other phrases. As an example, the word "twice" is always replaced by "2 times." The purpose of such substitutions is to reduce the number of linguistic forms by replacing many different special constructions with fewer, more general, ones.

Operators are phrases that denote the arithmetic operations STUDENT is familiar with (that is, addition, subtraction, multiplication, division, and exponentiation), or combinations of these operations (such as "the sum of x and the product of y and z"), or, finally, the relation of equality.

Some of the linguistic forms that can be used to express arithmetic operations and the equality relation are:

SUMMARY OF LINGUISTIC FORMS TO EXPRESS ARITHMETIC FUNCTIONS
AND THE EQUALITY RELATION

$x = y$	x is y; x equals y; x is equal to y
$x + y$	x plus y; the sum of x and y; x more than y
$x - y$	x minus y; the difference between x and y; y less than x
$x * y$	x times y; x multiplied by y; x of y (if x is a number)
x/y	x divided by y; x per y

The third category of words and phrases distinguished by STUDENT is that of variables. A variable is a sequence of words that STUDENT takes to represent an unknown of the problem which is presented. STUDENT makes no attempt whatever to understand the meaning of a variable; all it cares about is to identify which phrases are variables, and which phrases denote the same unknown. If STUDENT can solve a word problem in which the variable "first number" occurs, it could solve the similar word problem in which every occurrence of "first number" has been changed to some nonsense word like "fgh." The fact that STUDENT does not attempt to understand anything about the nature of a given variable makes it obvious that STUDENT's ability to recognize different phrases which denote the same unknown is limited. As an example, STUDENT might not recognize that "the number of years elapsed since Ann's birth" denotes the same unknown as "Ann's age."

The main idea of how STUDENT recognizes algebraic equations in verbal disguise can now be described as follows. After all substitutors have been replaced, STUDENT looks for occurrences of operators by matching phrases in the input sentences against the operator phrases it has stored in a catalogue. Then it identifies the sequences of words between the operators as variables and decides which sequences denote the same unknown (as indicated before, STUDENT will recognize two phrases as denoting the same unknown only if they are identical or differ in a minor way).

If the input sentences were of the kind STUDENT can understand, at this stage STUDENT will have constructed a set of algebraic equations in a formal mathematical notation. If it knows how to solve these equations, it proceeds to do so and prints out the results. If not, which happens if some equations are nonlinear or if there were fewer equations than unknowns, STUDENT asks "Do you know any more relationships among these variables?", and if this is answered with no, it prints "I can't solve this problem."

It is remarkable that this approach allows STUDENT to perform as well as it does. Of course, after one has understood how STUDENT works, it is easy to construct examples on which it will fail. For example, in a problem that involves the variable "the number of times I saw her," STUDENT will promptly identify "times" as the multiplication operator, and hence assume there are two variables involved, "the number of" and "I saw her."

However, the problems STUDENT has solved seem to indicate that, within its narrow algebraic scope, it rivals the average high school student. This fact lends some weight to the conclusion of D. G. Bobrow, who wrote STUDENT: "I think we are far from writing a program that can understand all, or even a very large segment, of English. However, within its narrow field of competence, STUDENT has demonstrated that 'understanding' machines can be built."

6.2. CAN A MACHINE REPRODUCE ITSELF?

Self-reproduction, the ability of an organism to produce others like itself, has sometimes been considered to be a prerogative of living organisms, and a popular indicator of the distinction between living and nonliving matter. Recent achievements in laboratory synthesis of complex replicating molecules have been called creation of "life in the test tube," further attesting to the popular view that self-reproduction and life are inseparable concepts. The fact that there are, however, several phenomena occurring in nature involving nonliving things that closely resemble self-reproduction (e.g., the growth of certain mineral crystals) suggests that the distinction between things which reproduce themselves and things which do not need not be the same as between living and not living.

There is a seemingly plausible argument which suggests that it might be impossible to construct a self-reproducing machine: If a machine A builds a machine B, then surely A must contain a complete description of B; and presumably A must contain some other parts as well, such as a control device to interpret this description, mechanical arms to perform the construction, and so on. Thus it appears that A would have to be more complex than B, and that no machine could build another of its own complexity.

Experience with machines we know seems to confirm this. For example, a machine on an assembly line that does nothing but manufacture screws must be significantly more complex than a screw. Despite the intuitive appeal which this argument might have, it is misleading, as the following discussion will show. While we have a great deal of experience with relatively simple machines, our intuition about what machines can do in principle (particularly if they are complex) is simply not very good.

John von Neumann was interested in such issues and studied them quite thoroughly in the early 1950s. He suggested that the ability of an organism to reproduce itself was indicative of the complexity of the organism, and concerned himself with the study of "self-reproducing automata," which could be completely and rigorously defined, rather than living organisms, which are so highly complicated that no one understands them completely.

Von Neumann considered several different models of machines in this study, two of which we shall discuss here: first, a mechanical device, built in a rather conventional way as an assembly of elementary parts; and, second, an abstract algebraic model. The first model has the advantage of having great intuitive appeal to our experience with mechanical devices; hence, the construction of a self-reproducing machine is easily understood in its main ideas. It has the disadvantage that a detailed construction, say in the form of a blueprint of the machine, would be very complicated and time consuming. The second model tries to circumvent this need for excessive concern with detail by presenting the idea of self-reproduction in a simplified setting. It then becomes feasible to carry out a detailed design of a self-reproducing machine; Von Neumann did so in a manuscript of approximately 200 pages. A disadvantage of the second model for expository purposes is that it takes a considerable amount of thought to understand precisely in what sense this algebraic model captures the notion of machine self-reproduction. Hence we shall devote only a few words to this model, and then outline the construction of a self-reproducing machine in von Neumann's first model.

In the algebraic model, usually called the *cellular* or *tessellation model*, space is represented as a two-dimensional infinite rectangular array of identical cells, and time is assumed to progress in discrete steps. Every cell is itself a simple abstract machine, capable of assuming one of a finite number of states at any time instant, and able to exchange signals with its four immediate neighbor cells. There is a transition function that determines for each cell its behavior in time, prescribing a state change from one instant to the next depending only on the present states of the cell and its four neighbors. One of the possible states of a cell is called a *quiescent state*, and a cell in this state may be considered an empty cell. A machine, in this model, is identified with a finite geometric pattern of contiguous cells and a specification of a state for each cell in the pattern, and operates in the following manner. Prior to some initial time, t_0, all the cells in the infinite array are in the quiescent state. At time t_0 the cells in the geometric pattern are all set to their specified states; the others remain as they are. From then on, the operation is governed step by step by the transition function, determining individual state changes for each cell. One can then call a machine (a pattern of contiguous cells in certain states) self-reproducing if at some finite time after t_0 there are two finite patterns embedded in the infinite array, each identical to the original machine.

Let us now describe the first model of self-reproduction mentioned. Imagine a gigantic warehouse that contains a large, potentially unlimited supply of a finite number of machine parts. These are the "elementary parts" from which the self-reproducing machine is built and should be of sufficient variety that mobile, self-powered vehicles could be constructed (e.g., wheels, gears, beams, nuts and bolts, paper tape, and batteries). A precise description of this supply of parts is a matter of some concern, as will be seen later, but for the moment assume that a collection can be suitably defined. The question of the existence of a self-reproducing machine in this setting can be phrased as: "Does there exist, for a sufficiently rich inventory of elementary parts, a machine, built from these parts, that will roam around this warehouse, grabbing parts from bins and putting them together to build a copy of itself, a second identical machine, which could then proceed to construct a third copy, and so on?"

Here is the concern mentioned above: it is quite easy to define the problem away by an unreasonable choice of the elementary parts. For example, the question becomes trivial if one of the elementary parts is defined as a complete machine minus a battery, or something as simple as that; for then self-reproduction becomes nothing more than installation of the battery. One is left quite unsatisfied by such an argument, since it assumes for an elementary part essentially the same property it strives to establish for the whole machine. On the other hand, if one chooses the elementary parts too small or too simple, then the essential properties of the machine may become obscured by the details of construction with these parts. There is no rigorous set of rules to guide the definition of the elementary parts, but only common sense, and the criterion for the appropriateness of a particular choice should be whether or not the resulting problem is "interesting."

Von Neumann himself at one time described a group of about a dozen elementary parts that he thought could be used in building his machine, although he did not carry out the description in terms of these parts. He instead broke down the self-reproducing machine into three smaller machines, none of which by itself is capable of reproducing itself, and which are all quite plausible machines. These three machines are (1) U, a universal constructor; (2) D, a tape duplicator; and (3) C, a control automaton. We shall not describe the laborious construction of these machines from a collection of elementary parts, but instead we appeal to your intuition that such a collection and construction is possible in principle, and describe these machines and their operation directly.

The universal constructor, U, is the most critical part of von Neumann's model of a self-reproducing machine. U is a machine that, given a description, $d(M)$, of an arbitrary machine, M, built of the parts $P_1, \ldots, P_k$, will build a copy of the machine M from those parts. The analogy with a general-purpose (universal) computer, which, given sufficient memory, can be programmed to compute any "computable" function (see Section 6.3) convinces most people familiar with computers that such a machine could be built. A computer-controlled assembly-line manufacturing plant comes quite close to the machine U. A complete design of U (specification of the list of parts, of a description, and of a construction algorithm) would seem to be a major engineering effort, but well within the capabilities of modern technology.

The universal constructor U requires for its operation an unambiguous description, in a machine-readable form, of the machine it constructs. This creates another concern in the choice of the class of elementary parts; that is, the parts must be so chosen as to guarantee that there exists a *finite* description of any machine M which can be built with them. This imposes no great restriction in practice, since it is perfectly reasonable to assume a set of parts with the property that if two parts can be joined, at all, they can be joined in only a finite number of ways.

There is no particular difficulty, once a suitable class of parts is chosen, in designing a formal language to describe the parts and the ways they may be interconnected. In the discussion to follow, we shall say that a description $d(M)$ of a machine M is encoded in the form of a punched paper tape, although any machine-readable description would suffice.

The other two component machines are much simpler than U. The tape duplicator, D, is simply a machine that will read a punched paper tape and make a copy of it by punching an identical pattern of holes in a blank tape. The control automaton, C, is a simple machine which can control the operation of U and D, that is, turn them on or off, insert tapes into them, and so on. The specific function of C will be clear with the description of the behavior of the composite machine $U + D + C$ that follows.

Let M be an arbitrary machine and $d(M)$ its description in terms of $P_1, \ldots, P_k$ punched into a paper tape; let $d(M)$ be attached to the machine $U + D + C$. First, the control machine C inserts the tape $d(M)$ into the duplicator D, resulting in two copies of $d(M)$. Then C inserts one copy of $d(M)$ into the universal constructor U, which then constructs machine M from the description. Finally, C attaches the other copy of $d(M)$ to the newly built M and reattaches the original $d(M)$ to $U + D + C$. Thus $U + D + C + d(M)$ for an arbitrary machine M constructs $M + d(M)$. For self-reproduction, then, all that is necessary is to attach to $U + D + C$ the tape $d(U + D + C)$; for then $U + D + C + d(U + D + C)$ constructs precisely a copy of itself.

This completes our discussion of von Neumann's models of self-reproducing machines. We have seen a machine that operates in an interesting environment (one in which the elementary parts can be as simple and familiar as those we encounter in everyday life) and yet itself has the unusual capacity for self-reproduction, which in our experience we have seen only in living beings. Thus the original question of this section has been answered affirmatively.

6.3. SOME THINGS MACHINES CANNOT DO:
LOGICAL LIMITATIONS

The topics we have considered in the previous sections have in common that they investigate aspects of the comparison between men and machines; they report on the extent to which computing machinery can be made to mimic, in some sense, the behavior of living organisms. In both sections the conclusion is that sufficiently complex machines could be made, in time, to mimic such behavior in certain limited cases, like conversation in natural languages, self-reproduction, and the like. In this section we shall no longer be concerned with constructing such intelligent machines or with the mind–machine comparison, but rather we shall look at the investigation of the basic nature of algorithmic, machine-like computations themselves, be they carried out by machines or by people.

The notion of an algorithmic computation is an informal, intuitive concept that is not immediately suited to a rigorous logical treatment. However, the problem has been successfully studied by the establishment of formal characterizations that attempt to capture the essence of algorithmic computation in a mathematically rigorous theory, and although it is impossible to establish rigorously that such formal characterizations accurately mirror the informal notion, it is generally accepted that they do. (The hypothesis that they do is commonly called *Church's thesis*, after the famous logician Alonzo Church.) The formal theory of computability, or theory of recursive functions, studies the question: What computations can be carried out by a machine if all limitations of a practical nature, such as the time and memory size required, are ignored? Some of the main results of this theory are negative in nature; they assert that certain computations cannot be carried out (or certain problems cannot be solved, or certain questions cannot be answered) by *any* machine whatsoever. It is remarkable that the basic results in this area predate the first electronic digital computers by more than a decade.

The theory of computability is usually presented in such a way that only computations on natural numbers are considered. This simplifies the presentation a great deal, but is no serious restriction, because any data that is presented to a digital computer or is output by a digital computer, by the fact that it is in digital form, can always be considered to be a single integer, albeit perhaps a very large one.

To develop computability theory rigorously, one should define precisely the type of machine considered for carrying out the computations. To make this definition short and easily understandable, it is convenient, in theoretical studies, to choose conceptually simple models of computing machines; the classical example is a *Turing machine* in which a finite control automaton using a finite number of symbols can read and write one symbol at a time on a potentially infinite tape. However, the resulting gain in the simplicity of the foundations is partly offset by the need to justify that these machines are, despite their apparent simplicity, *universal*, that is, powerful enough to simulate any other computer.

For the purpose of giving an expository presentation of a few results, rather than a rigorous development of the theory of computability, we prefer to choose as our model of a computing machine one that can be programmed in a language similar to some of the currently existing high-level programming languages. No more specific description of the machine will be given, but the reader can, if he wishes, visualize any general-purpose computer and any procedure-oriented high-level programming language with which he is familiar, and the discussion will apply. The notation we shall use will resemble several existing programming languages, but is not intended to be faithful to any particular one; it has been chosen only to make the few programs we write as self-explanatory as possible.

Before we can state some of the results we are aiming at, we have to clarify two notions: a special type of program, which we shall call *procedures*, that can be written in our language, and the *functions computed by such procedures*. This is best done by examples. The usual convention regarding integer division in a computer is that any fractional part of the quotient is dropped and the result truncated to the next lower integer. Under this convention the following procedure

$$G: \text{PROCEDURE } (N,M)$$
$$Y = N - (N/M)*M$$
$$\text{RETURN}(Y)$$
$$\text{END } G$$

will compute the function

$$g(n, m) = \text{residue of } n \text{ modulo } m$$

for any pair of positive integers (n, m).

The procedure G has two arguments, or input variables, N and M, and one output variable, or result, Y. For technical reasons, it will be convenient later on to deal only with procedures that have precisely one input and one output variable. However, since we are interested in procedures that compute functions of more than one argument, we have to agree on a method of encoding the n argument values $x_1, x_2, \ldots, x_n$ of a function f of n arguments into a single integer, which can be assigned to the input variable X of a procedure F that computes f. An encoding that is easily seen to be one to one is the following *Gödel numbering* scheme:

Assign to an n-tuple $x_1, x_2, \ldots, x_n$ of natural numbers the integer $\langle x_1, x_2, \ldots, x_n \rangle = p_1^{x_1} p_2^{x_2} \cdots p_n^{x_n}$, where p_i is the ith prime ($p_1 = 2$, $p_2 = 3, p_3 = 5$, etc.).

It is clear that any procedure F of n input variables, which computes a function $f(x_1, \ldots, x_n)$, can be modified to yield a procedure F1 of one input variable X, which computes the same function f under the convention that X is assigned the Gödel number $\langle x_1, x_2, \ldots, x_n \rangle$ of the n argument values $x_1, x_2, \ldots, x_n$. F1 can be obtained by preceding the body of F by some statements that decode $\langle x_1, x_2, \ldots, x_n \rangle$ to yield the n individual numbers $x_1, x_2, \ldots, x_n$.

As an illustration, the following procedure G1, when presented with an integer k of the form $2^n 3^m$, proceeds to decode k and then computes the residue of n modulo m just as the previous procedure G did.

```
G1: PROCEDURE(X)
    N = 0                              N and M are used to count how
    M = 0                              many factors of 2 and 3, respec-
                                       tively, X has;
    LOOP2: DO WHILE (X = (X/2)*2)
        X = X/2                        while X is even, remove one
        N = N + 1                      factor of 2 and increment the
                                       counter N;
    END LOOP2
    LOOP3: DO WHILE (X > 1)
        X = X/3                        remove one factor of 3 and incre-
        M = M + 1                      ment the counter M;
    END LOOP3
    Y = N - (N/M)*M
    RETURN(Y)
END G1
```

When we wish to emphasize the fact that the procedure G1 computes a function of two arguments (N and M) under an appropriate coding, we use the notation

$$\text{G1: PROCEDURE}(\langle N, M \rangle).$$

We shall use such a notation in later examples.

After these preliminaries, let us now consider the first question that arises at this point, that is, whether or not every function that maps natural numbers into natural numbers can be computed by some procedure written in a given language. The answer is no, and it can be proved easily by the following argument that there are not enough programs to compute all such functions.

Any programming language uses a finite set of basic characters or symbols from which the program statements are built (the standard character set found on a keypunch keyboard is an example), and any program in the language is a finite string of such characters. Thus the class of all possible procedures in a given programming language is at most denumerably infinite, that is, can be put in a one-to-one correspondence with the set of natural numbers. The set of all functions that map natural numbers to natural numbers, however, is nondenumerable; it cannot be put in a one-to-one correspondence with the set of natural numbers. This now well-known fact was proved in the late nineteenth century by Cantor, using the following *diago-nalization technique*.

Assume that the set of all functions from natural numbers to natural numbers was denumerable so that we could speak of the first function, the second function, and, in general, the *i*th function f_i. Now define a function g as follows:

$$g(i) = f_i(i) + 1.$$

g is a function from natural numbers to natural numbers, which, by its very definition, is different from every function f_i, and hence we have a contradiction to our assumption that the set of all functions could be enumerated.

By the above, the computable functions are a denumerable subset of the set of all functions on the natural numbers. Since we have just shown that the latter contains uncountably many functions, there are functions among them (indeed, most of them) that are not computable by any procedure.

This result is not particularly satisfying, because it gives no insight into the nature of the noncomputable functions. However, it is possible to give explicit examples of functions that are noncomputable, and provably so. Consider, for example, the following problem. Suppose that among the procedures in our language there was a special, "very smart" procedure that could decide for an arbitrary procedure P_k and arbitrary input n whether or not the computation of $P_k(n)$ would ever come to a halt. Let H be such a procedure:

```
H: PROCEDURE(⟨K,N⟩)
     ┌Very smart program to ┐
     │generate and examine  │
     │P_K and decide whether│
     │it will halt on input │
     └N.                    ┘
     IF P_K(N) would halt RETURN(1)
          ELSE RETURN(0)
   END H
```

H would be called a *decision procedure* for the *halting problem* for programs. We show by a contradiction that H cannot exist. Define the procedure

```
G: PROCEDURE(X)
     L: IF H(⟨X,X⟩) = 1 GO TO L
             ELSE RETURN(X)
   END G
```

If H is a legal procedure, then so is G, and hence G has an index x_0 in the listing of all legal procedures; that is,

$$G = P_{x_0}.$$

Now consider the computation $H(\langle x_0, x_0 \rangle)$. By its definition

$$H(\langle x_0, x_0 \rangle) = \begin{cases} 1 & \text{if } P_{x_0}(x_0) \text{ halts,} \\ 0 & \text{if } P_{x_0}(x_0) \text{ does not halt;} \end{cases}$$

but if $H(\langle x_0, x_0 \rangle) = 1$, then by the definition of G, $G(x_0)$ loops forever; that is, $P_{x_0}(x_0)$ runs forever. But then $H(\langle x_0, x_0 \rangle) = 0$, a contradiction. Similarly, if $H(\langle x_0, x_0 \rangle) = 0$, then $G(x_0)$ halts and returns x_0; that is, $P_{x_0}(x_0)$ halts, and $H(\langle x_0, x_0 \rangle) = 1$, again a contradiction. Thus we have a program G and an input x_0 for which H fails to compute the correct answer. We conclude that H cannot be a legal procedure. The result of this argument can be stated as

Theorem

There is no procedure that can decide for an arbitrary procedure P and input n whether the computation $P(n)$ will halt.

The same proof establishes a slightly stronger result, which we shall need later on: there is no decision procedure even for the more restricted (diagonal) halting problem which asks whether procedure P_k halts with input k.

One property of the denumerable set of procedures in a programming language that has far-reaching implications is that they can be numbered in a particularly nice way: For any programming language there is an algorithm (part of the compiler for the language) to determine of a given string whether it is in fact a legal program in the language. It is also clear that there is an algorithm which recognizes all legal procedures with exactly one input variable, and these are the ones of interest for the following discussion. If we were to couple this latter algorithm with an algorithm to generate all possible finite strings of basic characters, we could get an algorithm that generates a list of all procedures in the language which have precisely one input variable and one output variable. This list provides a numbering of these procedures that is effective in the following sense. Suppose that we let P_n be the nth procedure to be listed by the algorithm. Then given any number k, we need only to generate the first k procedures in the list to obtain a copy of the procedure P_k; conversely, given any procedure P, we need only generate the list, counting as we go, until P itself is generated, to obtain the index k such that $P = P_k$.

One striking implication of this effectiveness of the numbering of all procedures is the existence of universal programs. This result says that there is a procedure U which, given the index k of an arbitrary program P_k and given m, will simulate the computation of P_k on m.

> U: PROCEDURE($\langle$k,m$\rangle$)
> $\begin{bmatrix} \text{Program to generate the } k\text{th} \\ \text{program in the list of programs} \\ \text{and to execute } P_k(m). \end{bmatrix}$
> RETURN(P_k(m))
> END U

Thus one such universal program can compute any computable function.

The unsolvability of the halting problem is not an isolated instance. There are many mathematical problems whose general solution would be useful, but that have been proved to be algorithmically unsolvable. One of the most prominent among these is the problem of deciding, for an arbitrary assertion in a sufficiently rich mathematical theory, whether or not it is a theorem; that is, whether or not it follows from the axioms of the theory.

Let us show the unsolvability of one more problem, that of deciding for two arbitrary procedures P_m and P_n whether or not they are equivalent in the sense that they compute the same function. The following argument will show that for any class of procedures large enough to include the effective numbering algorithm and universal procedures, there can be no procedure for deciding the equivalence of programs.

Suppose that such a procedure did exist; call it E.

> E: PROCEDURE($\langle$m,n$\rangle$)
> $\begin{bmatrix} \text{Very smart program to generate} \\ P_m \text{ and } P_n \text{ and determine whether} \\ \text{or not they define the same function.} \end{bmatrix}$
> IF yes RETURN(1)
>   ELSE RETURN(0)
> END E

We shall reduce the unsolvability of this *equivalence problem* to that of the halting problem. That is, we shall show that the procedure E, along with some others whose existence is unquestionable, would be sufficient to solve a case of the halting problem which the previous argument showed was impossible, and hence that the assumption that E exists leads to a contradiction.

To relate the equivalence problem more closely to the halting problem, choose some simple function such as the zero function that maps all natural numbers to zero, and choose a fixed procedure that computes this function, such as the procedure

ZERO: PROCEDURE(X)
RETURN(0)
END ZERO.

A simple modification of this procedure now yields the following infinite set of procedures whose properties reveal the close connection between the equivalence problem and the halting problem.

Z_1: PROCEDURE(X)
CALL $P_1(1)$
RETURN(0)
END Z_1

.
.
.

Z_n: PROCEDURE(X)
CALL $P_n(n)$
RETURN(0)
END Z_n

.
.
.

The "CALL $P_n(n)$" statement of the procedure Z_n is to be regarded as an abbreviation for the statements of the universal procedure U, where the constant n replaces each occurrence of k and m. With this understanding, it is clear that these procedures can be written in the language we are considering.

Each procedure Z_n computes a trivial function of one variable x. For each value of x, Z_n simulates the computation $P_n(n)$ and, *if it halts*, returns the value zero. Thus each Z_n computes the zero function if and only if P_n halts on input n; if $P_n(n)$ fails to halt, then Z_n also fails to halt for any input x.

The set of procedures Z_1, Z_2, . . . is a subset of the set of all procedures P_1, P_2, . . . which can be written in the language we are using. Thus every Z_n has an index, call it I_n, in the list of all procedures (so that Z_n and P_{I_n} are the same procedure). Since we have assumed a class of programs that contains a program which implements the effective enumeration algorithm, it is possible to write a procedure INDEX, which, given an input n, does the following two things: first, it generates procedure Z_n, and, second, it finds the index I_n of Z_n.

INDEX: PROCEDURE(n)
⌈Generate procedure Z_n⌉
⌊and find its index I_n.⌋
RETURN(I_n)
END INDEX

Now, since we assumed the procedure E is capable of deciding the equivalence of two arbitrary procedures, it follows in particular that E can decide, for any n, whether or not Z_n is equivalent to the special procedure ZERO, provided only that E is executed with the index I_n of Z_n and with the index (call it I_0) of ZERO.

$$E(\langle I_0, I_n \rangle) = \begin{cases} 1 & \text{iff } Z_n \text{ computes the zero function iff } P_n(n) \text{ halts}; \\ 0 & \text{iff } Z_n \text{ does not compute the zero function, iff} \\ & P_n(n) \text{ does not halt.} \end{cases}$$

Thus using the alleged procedure E and the procedure INDEX described earlier, we can write the following procedure H', which will decide for every natural number n whether P_n halts with input n.

```
H': PROCEDURE(n)
    CALL INDEX(n)
    CALL E(⟨I₀,INDEX(n)⟩)
    RETURN(E(⟨I₀,INDEX(n)⟩))
    END H'
```

Since we showed that such a procedure H' cannot exist, we conclude that E cannot exist.

The two results we have seen in this section are typical of the unsolvability results that are part of computability theory. These have profound implications for those with an interest in computing; they are absolutes in the sense that they identify impossible tasks on which much effort could be wasted. Recently, researchers have begun to study computability under certain restrictive assumptions about the amount of computational resources available: for example, what can be done if some bound is placed on the length of the computation or the amount of memory available. This is an area of great current activity that promises even wider implications for the field of computation. We shall not discuss this further here but refer the interested reader to the references cited in the next section.

6.4. REMARKS AND REFERENCES

Von Kempelen's fake chess automaton is mentioned frequently in the literature. It has been said to have been partly responsible for the development of interest in the game in America in the early part of the nineteenth century. Its interesting history is sketched in

HAGEDORN, R. K. *Benjamin Franklin and Chess in Early America*, University of Pennsylvania Press, Philadelphia, 1958.

Interest in chess-playing programs has been active for as long as digital computers have been in existence. Their development could well have served as another illustration of how computers have been programmed to perform at a nontrivial level in an activity commonly regarded as having a high intellectual content. See Section 3.4 for references in this subject.

Turing's original discussion of the question "Can a machine think?" appeared in

TURING, A. M. "Computing Machinery and Intelligence," *Mind*,
9 (1950), 433–460,

and is reprinted in Volume 4 of *The World of Mathematics* edited by J. R. Newman (Simon and Schuster, New York, 1956). This paper, which is generally considered a classic in the literature, was one of the first contributions to a lively discussion of the mind–machine problem. A nice anthology of philosophical papers on this topic that includes it, as well as some criticism of it, helps to put the question of machine thinking into perspective; see

ANDERSON, A. R. (ed.). *Minds and Machines*, Prentice-Hall, Englewood Cliffs, N.J., 1964.

A description of several important programs for answering questions posed in natural language can be found in

MINSKY, M. (ed.). *Semantic Information Processing*, MIT Press, Cambridge, Mass., 1968.

Chapter 3 in this book describes the algebra problem solver, STUDENT, which was developed in a Ph.D. thesis at MIT by D. G. Bobrow ("Natural Language Input for a Computer Problem-Solving System").

The conversational program, ELIZA, was first described in

WEIZENBAUM, J. "ELIZA—A Computer Program for the Study of Natural Language Communication Between Man and Machine," *Comm. ACM*, 9 (1966), 36–45,

and again later, in

WEIZENBAUM, J. "Contextual Understanding by Computers," *Comm. ACM*, 10 (1967), 474–480.

These topics and many others regarding computer simulation of human behavior are also discussed in

APTER, M. J. *The Computer Simulation of Behavior*, Harper & Row, New York, 1970.

John von Neumann's discussion of self-reproducing machines was part of a paper outlining his concept of a systematic theory of automata. It appeared under the title "The General and Logical Theory of Automata" in

JEFFRESS, L. A. *Cerebral Mechanisms in Behavior—The Hixon Symposium*, 1–41, Wiley, New York, 1951.

It is also reprinted in *The World of Mathematics*, Vol. 4, mentioned earlier, and in

> TAUB, A. H. (ed.). *John von Neumann—Collected Works*, Vol. V, pp. 288–328, Macmillan, New York, 1963.

Further understanding of this type of self-reproduction can be gained by solving Exercise 2. In connection with this, see

> BRATLEY, P., and J. MILLO. "Computer Recreations," *Software-Practice and Experience, 2* (1972), 397–400.

The tessellation model of self-reproduction mentioned in the text was described in detail by von Neumann in a manuscript that was begun in 1952 and was left incomplete at the time of his death. This manuscript was subsequently edited and completed by A. W. Burks, and published as part of

> VON NEUMANN, J. *Theory of Self-Reproducing Automata*, (edited and completed by Arthur W. Burks), University of Illinois Press, Urbana, 1966.

There continues to be new research activity in this area, and a recent summary can be found in

> BURKS, A. W. (ed.). *Essays on Cellular Automata*, University of Illinois Press, Urbana, 1970.

This tessellation model is nicely embodied in the Game of Life as described in the Mathematical Games section of the October 1970 and February 1971 issues of *Scientific American*.

Section 6.3 on the logical limitations of machines is a brief introduction to the theory of computability, or the theory of recursive functions, whose foundation was laid in the 1930s by such logicians as Post, Turing, Church, and Kleene. Several of the important early papers are reprinted in

> DAVIS, M. *The Undecidable*, Raven Press, Hewlett, N.Y., 1965.

A recent textbook that provides a readable introduction to computability particularly accessible to the computer scientist is

> MINSKY, M. *Computation: Finite and Infinite Machines*, Prentice-Hall, Englewood Cliffs, N.J., 1967.

One major result proved in this text is the equivalence of several bases for computability, from the classical Turing machine, through the general recursive formalism to programs written for a simple computer model that is relatively close in concept to modern machines. For an in-depth treatment of the mathematical theory of recursive functions and an extensive bibliography, the reader might consult

> ROGERS, H. *Theory of Recursive Functions and Effective Computability*, McGraw-Hill, New York, 1967.

6.5. EXERCISES

1. Design and write a conversational computer program that answers questions in the manner in which an evasive diplomat might respond to questioning by newspaper reporters. He would probably make frequent use of such stock expressions as "I DON'T KNOW," "NO COMMENT," or "IT IS TOO EARLY TO SAY," which would serve as answers to almost any question. However, tradition would force him to frequently give answers that are specifically related to the question posed, even though they do not answer the question. For example, "WILL YOUR GOVERNMENT INCREASE MILITARY EXPENDITURES?" might be answered with "MY GOVERN-MENT IS CURRENTLY RECONSIDERING ITS COMMITMENTS." This response might be composed by combining the stock phrase "IS CUR-RENTLY RECONSIDERING ITS COMMITMENT" with the phrase "MY GOVERNMENT" obtained by a transformation from "YOUR GOVERNMENT."

2. In whatever programming language you desire, write a self-reproducing program. Your program should print out an *exact* copy of itself. Any data read by your program should be considered to be part of the program.

3. One well-known "pairing function" to encode two positive integers into one is given by

$$\langle x, y \rangle = \frac{(x + y - 1)(x + y - 2)}{2} + y.$$

Prove that the function $\langle x, y \rangle$ is not only one to one, but also onto; that is, every integer i is the encoding of some pair (x, y). Find the inverse functions, l and r, which have the property that

$$l(\langle x, y \rangle) = x \quad \text{and} \quad r(\langle x, y \rangle) = y.$$

4. By using the pairing function of the previous exercise, show that rational numbers (positive *and* negative) *are* countable.

5. Using a diagonalization argument similar to that in Section 6.3, show that the real numbers *are not* countable.

6. Suppose the halting problem was solvable; are there still unsolvable problems? In other words, suppose that we assume the existence of an *oracle* (some mystical being, not a procedure) who can always supply the correct answer to the question "Does P_k halt on input n?" Given this oracle, are there still problems that are unsolvable?

7. Is there a procedure P_k which, on input n, returns the value 1 if P_n halts on *all* inputs, and returns the value 0 otherwise?

8. Is there a procedure which, on input n, returns the value 1 if P_n halts on input n and does not halt otherwise?

INDEX

247